Schriftenreihe der Professur für Molekulare Lebensmitteltechnologie

Band 10

Hemisynthesis of phenolic metabolites

Dissertation

zur Erlangung des Doktorgrades (Dr. rer. nat.)

der

Mathematisch-Naturwissenschaftlichen Fakultät

der

Rheinischen Friedrich-Wilhelms-Universität Bonn

Vorgelegt von

Sarah Monika Straßmann geb. Schmitt

aus

Wiesbaden

Bonn 2021

Bibliografische Information der Deutschen Nationalbibliothek

Die Deutsche Nationalbibliothek verzeichnet diese Publikation in der Deutschen Nationalbibliografie; detaillierte bibliographische Daten sind im Internet über http://dnb.d-nb.de abrufbar.

1. Aufl. - Göttingen: Cuvillier, 2021

Zugl.: Bonn, Univ., Diss., 2021

Angefertigt mit Genehmigung der Mathematisch-Naturwissenschaftlichen Fakultät der Rheinischen Friedrich-Wilhelms-Universität Bonn

1. Gutachter: Prof. Dr. Andreas Schieber

IEL – Molekulare Lebensmitteltechnologie, Universität Bonn

2. Gutachter: Prof. Dr. Matthias Wüst

IEL – Lebensmittelchemie, Universität Bonn

Tag der Promotion: 26. Oktober 2021

Erscheinungsjahr: 2021

Nonnenstieg 8, 37075 Göttingen

Telefon: 0551-54724-0

Telefax: 0551-54724-21

www.cuvillier.de

1. Auflage, 2021

Gedruckt auf umweltfreundlichem, säurefreiem Papier aus nachhaltiger Forstwirtschaft.

ISBN 978-3-7369-7525-5

eISBN 978-3-7369-6525-6

Table of contents

Preliminary remarks

List of abbreviations

ADME	Absorption, distribution, metabolism and excretion
ATP	Adenosine triphosphate
CCS	Collision cross section
COMT	Catechol-*O*-methyl transferase
COX	Cyclooxygenase
DNA	Deoxyribonucleic acid
ESI	Electrospray ionization
HRMS	High resolution mass spectrometry
HSA	Human serum albumin
IMS	Ion mobility spectrometry
LC	Liquid chromatography
LDL	Low density lipoprotein
MS	Mass spectrometry
MT	Methyl transferase
NMR	Nuclear magnetic resonance
OH	Hydroxyl group
PAPS	3'-phosphoadenosine-5'-phosphosulfate
QTOF	Quadrupole time of flight mass spectrometry
ROS	Reactive oxygen species
SGLT	Sodium-glucose linked transporter
SULT/ST	Sulfotransferase
UDP	Uridine diphosphate
UDPGA	Uridine diphosphate glucuronic acid
UGT	Uridine 5'-diphospho-glucuronosyltransferase
UHPLC	Ultra high-performance liquid chromatography
SPE	Solid phase extraction

List of publications

Schmitt, S., Tratzka, S., Schieber, A., Passon, M. Hemisynthesis of Anthocyanin Phase II Metabolites by Porcine Liver Enzymes. *Journal of Agricultural and Food Chemistry* **2019,** 67 (22), 6177-6189.

Straßmann, S.; Passon, M.; Schieber, A. Chemical Hemisynthesis of Sulfated Cyanidin-3-*O*-Glucoside and Cyanidin Metabolites. *Molecules* **2021**, *26*, 2146.

Straßmann, S.; Brehmer, T.; Passon, M.; Schieber, A. Methylation of Cyanidin-3-*O*-Glucoside with Dimethyl Carbonate. *Molecules* **2021**, *26*, 1342.

Conferences

Schmitt, S., Tratzka, S., Schieber, A., Passon, M. Semisynthese von Anthocyan-Phase-II-Metaboliten durch Schweineleberenzyme. 47. Deutscher Lebensmittelchemikertag in Berlin, Germany, September 17-19, 2018, *Lebensmittelchemie* **2019.** [Poster]

Schmitt, S., Schieber, A., Passon, M. Hemisynthesis of anthocyanin phase II metabolites and their characterization by LC-ESI-IMS-QTOF mass spectrometry. 13th World Congress on Polyphenol Applications, Valletta, Malta, September 30 – October 1, 2019, Abstracts of papers 2019, p. [Oral presentation]

Declaration of contribution as co-author

The contribution of the co-authors to the scientific papers presented in Chapter 2, 3 and 4 are as follows:

Prof. Dr. Andreas Schieber proofread all manuscripts and contributed to the publication as the supervisor of the thesis.

Dr. Maike Passon advised on experimental work and supported interpretation and publication of the results. She proofread all manuscripts and was as corresponding author responsible for all formal aspects of the publications.

Sebastian Tratzka developed the method of the hemisynthesis with porcine liver enzymes

Tillman Brehmer optimized the chemical methylation and the analysis of the products. He further contributed to the interpretation of the results.

Chapter 1

General Introduction

1 Polyphenols

Polyphenolic substances are ubiquitous secondary plant metabolites. As a part of this group of phytochemicals, they belong quantitatively to the most important compound classes in nature. They are the second most abundant group of all organic compounds after sugars, with a total share of about 30% of the total biomass on earth.[1] Most of them are produced by plants and are known, among others, as plant pigments or antioxidants in foods. Other producers are fungi and microorganisms. However, these play a minor role in human intake, which is about 1 g per day.[2] Polyphenols have gained general awareness due to their presence and positive image in so-called super fruits and super foods, which has led to an increased uptake by the population.[3]

1.1 Properties

As secondary plant compounds, polyphenols are not involved in the primary metabolism of the plant, yet they are essential for the survival of the plant. Their ability to absorb light of different wavelengths serves the plant as protection against harmful UV-B rays. Especially plants that are exposed to high levels of solar radiation increasingly incorporate polyphenols into the epidermis.[4] There, the polyphenols act as a protective layer, thus preventing mutagenesis, cell death and the formation of oxygen radicals, for example.[5] A further function is the signal effect. As color pigments, it is mainly anthocyanins that give the flowers and fruits a strong color and attract animals that are responsible for plant reproduction by dispersing the plant seeds or pollinating the flowers.[6] Just as they can attract living organisms, polyphenols can also protect the plant from harmful organisms by forming physical or chemical barriers. They can repel predators by an anti-microbial effect, an astringent or bitter taste, or poor digestibility.[5,7,8] Like the antioxidant effect, the anti-microbial properties are based on the inhibition of harmful enzymes such as cellulases or pectinases, which can be released by microorganisms to destroy the plant cell wall.[5]

Furthermore, polyphenols that enter the soil after the plant has died can have a positive effect on decomposing microorganisms in the soil. The increased activity of the destructors results in a higher conversion of plant material and thus a higher nutrient density in the soil. Plants can also release polyphenols as allelochemicals to inhibit the growth of competing plants. The possible function of polyphenols as physiological regulators or messenger substances is also currently being discussed.[5]

1.2 Substance classes

The substance classes in which polyphenols occur are quite different and literature describes many possible ways of classifications.[9] A grouping based on their chemical structure seems appropriate: They can be divided into tannins with more than 15 carbon atoms, hydroxycinnamic acids with 9 carbon atoms, and flavonoids with 15 carbon atoms. All these can in turn be divided into subclasses. As the compounds relevant for this work are the class of flavonoids, the others are neglected here.

Flavonoids often occur connected to one or more sugar or acid moieties. The structure of flavonoids is based on the flavan framework, which consists of two benzene rings (A and B) connected by a heterocyclic pyran ring (C) and thus is a C_6-C_3-C_6-structure.[10] The class of flavonoids can be divided into subgroups such as, for example, flavanones like naringenin, flavonols like quercetin, or isoflavones like genistein. This division is created by varying the oxidation status of the C ring and the position of the B ring.[11] Since the focus of this work is on anthocyanins, only these will be discussed in the following sections.

2 Anthocyanins

Anthocyanins play a crucial role among the flavonoids due to their positive charge. Hence, they are also referred to as flavylium salts, with chloride being the most common counterion in plants.[12] They are colorful and occur in plants in the vacuoles of different cells as blossom dye or as pigment in other parts of the plant, for example in berries like black currant, raspberries, or blackberries.[6]

The six mostly distributed anthocyanidins in nature are shown in **Figure 1**. They are distinguished by different hydroxy and methoxy substitution patterns at the B ring.[13]

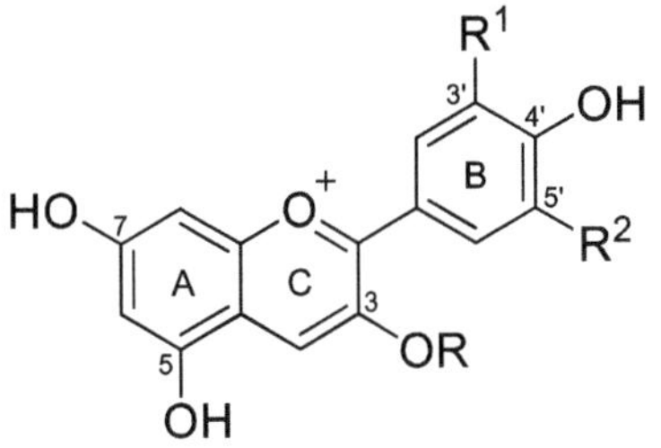

R^1	R^2	Anthocyanidin
H	H	Pelargonidin
H	OCH_3	Peonidin
H	OH	Cyanidin
OCH_3	OCH_3	Malvidin
OH	OCH_3	Petunidin
OH	OH	Delphinidin

Figure 1: Anthocyanin backbone with R = sugar. If R = H, it represents an anthocyanidin.

The aglycone is called anthocyanidin. Anthocyanidins are very unstable concerning light, high temperatures, and high pH values, which is why they are glycosylated in plants, esterified with organic or other phenolic acids or copigmented with other flavonoids.[14] Glycosylation on the C ring occurs most frequently. It is also observed on the two hydroxyl groups of the A ring, with position 5 being preferred.[15] The monosaccharides glucose, galactose, rhamnose, xylose and arabinose are the most common sugar residues. Di- and trisaccharides and multiple substitutions may also occur. Furthermore, glycosylation at some of the carbon atoms is possible.[6] In addition, the sugar residue may be esterified with aliphatic/aromatic acids.[13] Glycosylation is generally associated with an increase in the physicochemical stability and water solubility of the anthocyanins. The acylation of the sugar residues with cinnamic acid or aliphatic acids even promotes these properties.[16] According to He and Giusti, this increase in stability is attributed to the formation of intramolecular hydrogen bridge bonds.[3]

The focus of this work is on cyanidin, which is the most common anthocyanidin and cyanidin-3-*O*-glucoside as the most common anthocyanin.[6]

2.1 Chemical properties

Due to the substituted sugar residues, anthocyanins are water-soluble and are found in nature as pigments in the cell fluid of numerous plants, which has a slightly acidic environment. The color of the anthocyanins is exclusively due to the aglycone backbone, whose structural – and thus also color - appearance is strongly pH-dependent (**Figure 2**).[17]

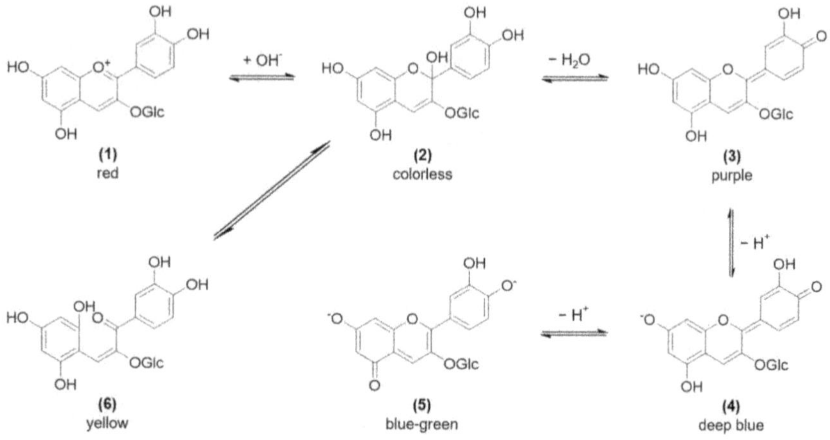

Figure 2: Structural formulae of cyanidin-3-*O*-glucoside at different pH values[17,18]

For anthocyanidins, the formation of chromenol (**Figure 2**, compound (2)) is already critical because they can then be further hydrolyzed via the intermediate stage of an α-diketone (**Figure 3**, compound (7)) to an aldehyde (**Figure 3**, compound (8)). This originates from the A ring and is identical for all aglycons. In addition, a carboxylic acid (**Figure 3**, compound (9)) is formed, which is characteristic for the respective anthocyanidin. The ring opening and thus the transition into the chalcone structure is an endothermic reaction. The decomposition process can be accelerated considerably, especially by increasing the temperature.[10]

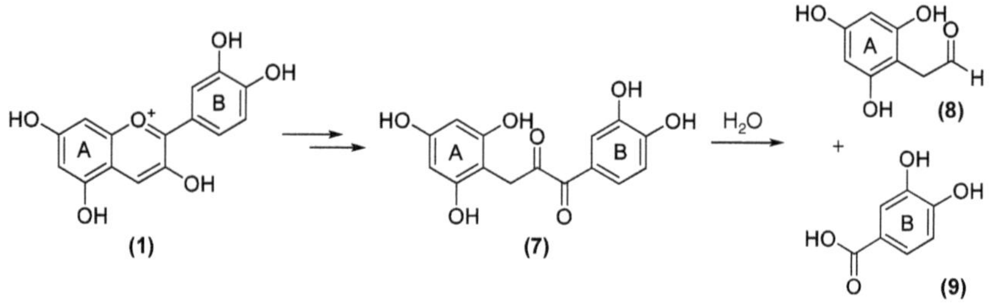

Figure 3: Decomposition of cyanidin at higher pH values[18]

2.2 Physical properties

Anthocyanins show characteristic UV absorption spectra, which can be used for identification and confirmation purposes during analysis. The spectra of anthocyanins, present in the form of flavylium cations (**Figure 2**, compound (1)), are characterized by two maxima depending on the respective solvent: One is located at 265 - 280 nm, while the other is in the visible range at 465 - 560 nm and is very pronounced. The UV absorption spectra of anthocyanins are pH-dependent: If no conjugated aromatic system is present at an pH value about 4.5 (structure (2), **Figure 2**), no maximum is observed at 465 - 560 nm.[19] This characteristic is illustrated in **Figure 4**.

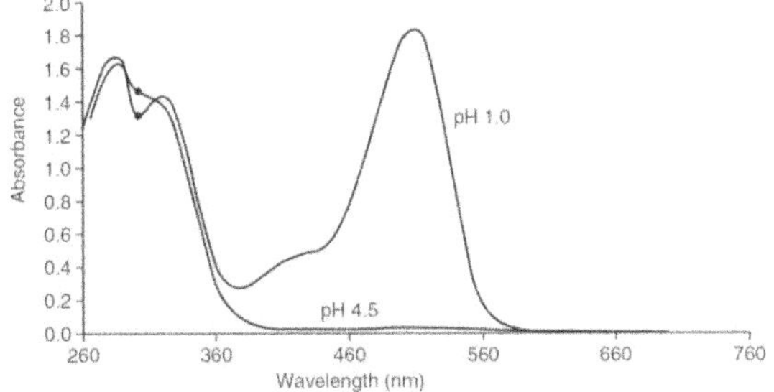

Figure 4: UV-Vis spectra of acylated pelargonidin-3-sophoroside-5-glucoside derivatives at pH 1.0 and 4.5.[19]

Structural differences can be identified using the UV spectrum. Thus, 3-monoglucosides can be differentiated from anthocyanins with a further sugar residue at the C-5 position. The monoglucosides have a shoulder at about 440 nm or before their second absorption maximum.[20] Conjugated groups as well as pH changes lead to shifts of the absorption maxima. Cruz and co-workers[17] investigated the changes of the absorption maximum of cyanidin-3-*O*-glucoside-7-*O*-glucuronide with increasing pH values. Moreover, certain conjugated groups may have additional absorption maxima, e.g., conjugation with acylated hydroxycinnamic acid leads to a third maximum at about 310 nm.[21] Glycosylation and methylation at the B ring cause a blue shift (shift to lower wavelengths, also: hypsochromic shift), while with increasing hydroxylation as well as inter- and intramolecular copigmentation a red shift (shift to higher wavelengths, also: bathochromic shift) is observed.[21] This knowledge found application in this work.

2.3 Physiological properties

It has been known for many centuries that a high proportion of vegetable food is healthy. In the past two decades, this traditional knowledge has been scientifically verified, with the result that the consumption of larger quantities of fruit and vegetables leads to a reduction in the incidence of cardiovascular diseases, cancer, osteoporosis, or diabetes.[3] Besides vitamins and minerals, mainly polyphenols have been identified as the cause of this positive health effect. Dietary antioxidants, such as anthocyanins, are able to increase the antioxidant capacity of the serum, which can provide low density lipoprotein (LDL) oxidation and thus prevent cardiovascular diseases such as arteriosclerosis.[3] Based on model experiments, it is also assumed that a diet rich in anthocyanins may have positive effects on the loss of cognitive abilities and neurodegenerative diseases with increasing age.[13] The neuroprotective efficacy is based in particular on the fact that anthocyanins are able to cross the blood-brain barrier of pigs and rats and are thus possibly also centrally effective.[22] Anthocyanins from berry fruits are also believed to have a positive effect on vision in several ways.[23] On the one hand, they improve night vision by increasing the generation of retinal pigments and, on the other hand, they are considered to help prevent glaucoma, retinitis pigmentosa and cataracts.[13] Anthocyanins are also said to possess an antidiabetic activity.[23] The antioxidant anthocyanins help to protect pancreatic cells from glucose-induced oxidative stress.[23] It is also believed that anthocyanin uptake might improve the function of adipocytes, thereby preventing metabolic syndrome or obesity.[3] It has been observed that all inflammatory parameters can effectively be reduced by a sufficiently high anthocyanin concentration.[3] Therefore anthocyanins are also supposed to have an anti-inflammatory effect[13]. The anticarcinogenic activity of anthocyanins is attributed to additive effects of several mechanisms. Among the mechanisms involved are antimutagenic activity, inhibition of oxidative DNA damage, inhibition of carcinogenic activation, stimulation of phase II enzymes for detoxification, stimulation of cell cycle arrest, inhibition of COX-2 enzymes, induction of apoptosis and anti-angiogenesis.[3] In particular, cyanidin and delphinidin have been shown to inhibit the growth of certain tumor cells, whereas the respective glucosides have been shown to be inactive in this regard.[6]

2.3.1 Antioxidativity and radical scavenging properties

Due to their numerous OH groups on aromatic rings, polyphenols are rich in electrons and therefore good nucleophiles. This means they can be easily oxidized and therefore used as antioxidants.[24] Phenols can release hydrogen ions. This is possible via radical or acid-base reactions. The aromatic ring can then stabilize the resulting compound well by resonance. Harmful biological radicals such as singlet oxygen, hydroxyl- and peroxyl radicals as well as nitrogen radicals, can thus be intercepted by them. Reactive oxygen species (ROS) can otherwise lead to oxidative damage to DNA, proteins, and lipids.[1]

There are various mechanisms by which anthocyanins unfold their antioxidative potential. For example, due to their electron-donaing properties, they can directly intercept free radicals and ROS[24]. Anthocyanins indirectly support the body's own antioxidative defense mechanisms by, for example, increasing the glutathione content in the cells[25], inhibit enzymes that are responsible for the formation of free radicals or regenerate other antioxidant substances such as α-tocopherol.[26–28]

2.3.2 Bioavailability

In order for a substance to exert an effect in the body, it needs to be absorbed into the blood and must be transported to its site of action. This process is called bioavailability. Knowledge of the bioavailability provides a crucial basis for assessing the physiological significance of bioactive food ingredients. Without any knowledge about the behavior of a substance in the body, it is almost impossible to make a statement about its efficacy. In order to investigate this, studies have to be carried out that explore every single way a molecule may pass in the body. The bioavailability of a nutrient can be strongly influenced by factors such as other food ingredients in the gastrointestinal tract, but also the change of the substance by metabolism.[29] In animal model studies, anthocyanins are detectable in the blood just a few minutes (6-20) after oral administration.[30] There are several assumptions about their uptake by humans and animals. One concludes that a part is already absorbed in the stomach as intact glycosides with the help of bilitranslocase and enters the bloodstream via the liver.[31] Another study shows absorption in the jejunum. It takes place actively via the sodium-dependent glucose co-transporter SGLT1.[32] In contrast, the aglycones can diffuse passively through the mucosa because they are more hydrophobic.[11] Consequently, anthocyanins can either be taken up via active mechanisms or they are absorbed passively by enzymes such as β-glucosidase converting them into their corresponding anthocyanidins.[33] Anthocyanins that have not been absorbed up to that point can be hydrolyzed by the colon microbiota. The resulting aglycons are broken down to phenolic acids, among other compounds, which can then pass through the intestinal wall.[34]

2.3.3 Absorption, distribution, metabolization and excretion (ADME)

It is known that, though their bioavailability is low, the human organism largely metabolizes dietary flavonoids to different conjugates that further appear in the bloodstream. Through this biotransformation the organism makes substances more hydrophilic to excrete them more efficiently. All of the following steps of the metabolism are highly dependent on various factors such as age, gender, genotype, microbiome and the composition and processing of the ingested food.[29] A recapitulation of the ADME process is illustrated in **Figure 5**. At each of the described points, it is possible that the body builds substances that can have a different effect than the original compounds. They can then, for example, change the effects described in section 2.3 or even cause them themselves. Therefore, more research concerning these metabolites is needed.

2.3.3.1 Mouth and stomach

Already in the mouth some polyphenols are deglycosylated and hydrolyzed by the oral microbiota. In the stomach, the polyphenols are further hydrolyzed by the acidic environment. Smaller phenolic acids can already be absorbed in the stomach, while some larger compounds reach the intestine completely unchanged.[35] Anthocyanins are the only group of flavonoids that are absorbed in this stage and are thus the first that reach the bloodstream.[36]

2.3.3.2 Small intestine

Monomeric flavonoids such as catechin can be absorbed unchanged by passive diffusion into the enterocytes.[37] In general, however, polyphenols are found in food as esters, glycosides or polymers. Due to their higher molecular weight, these cannot be absorbed directly, but are first depolymerised by the enzymes of the microflora. The glycosidic bond is cleaved and the aglycones are broken down to smaller molecules.[38–41] The smaller molecules are then also absorbed into the enterocytes by passive diffusion. Some of the glycosidically bound polyphenols are not degraded but taken up via D-glucose transporter proteins.[41] In the enterocytes, the absorbed phenolic compounds are glucuronidated by the UDP-glucuronosyltransferase and subsequently absorbed by the liver cells.[42,43] Some studies have shown that the absorbed aglycones in the enterocytes are not only glucuronidated but also partly methylated there and in the case of quercetin also sulfated.[44–46] Sulfation, glucuronidation and methylation are part of the biotransformation of xenobiotics in the human body. To excrete xenobiotics more effectively, water solubility and molecular weight are increased in two phases. In phase I a polar functional group is attached to the foreign substance by hydrolysis, oxidation, or reduction. These are catalyzed by enzymes like oxygenases, dehydrogenases, cytochrome P450 or esterases. In vitro experiments demonstrated that anthocyanins do not undergo phase I metabolism by cytochrome P450 enzymes.[47] In phase II, the water solubility and molecular weight of the foreign substance is further increased by sulfation and glucuronidation, among others. Methylation additionally reduces the pharmacological activity of foreign substances.[48–50] The phase II reactions are described in detail in section 2.3.4.

2.3.3.3 Colon (microbiome)

Only 5-15% of the flavonoids are absorbed in the small intestine; the majority reaches the large intestine. Here they are intensively metabolized by the intestinal bacteria. Dependent on the concentration, they are toxic to many intestinal bacteria and may thus change the spectrum of the intestinal biota. However, the microorganisms also produce new compounds like small phenolic acids or other aromatic catabolites, which in turn can be absorbed into the blood, be metabolized, and can have effects on health. This bidirectional relationship is the subject of current research. It is also investigated whether the catabolites may have a higher health-promoting effect than the phase II metabolites of intact anthocyanins.[29] To study this in humans, reference substances are urgently needed to investigate their effects in cell experiments or in vivo.

2.3.3.4 Liver

The liver plays an important role in the detoxification of the body. Xenobiotics and toxic substances can be converted and broken down. These can then be excreted through the urine or bile. Nutrient-rich blood from the intestine is supplied via the portal vein to the liver. Nutrients as well as drugs and other substances are absorbed and metabolized. Most of the compounds are already glucuronidated in the enterocytes, but some compounds also reach the liver unmetabolized and are glucuronidated there. A wide variety of metabolic reactions of unreacted, already metabolized or hydrolyzed compounds can take place here.[42,43,49] Methylation is carried out by *O*-methyltransferases, glucuronidation by UDP-glucuronosyltransferases and sulfation by sulfotransferases.[49,51] The enzymes and their reactions are described more in detail in section 2.3.4.

2.3.3.5 Blood

The metabolized phenols are now released from the stomach, intestine and liver into the blood, where they bind to plasma proteins such as human serum albumin (HSA).[14] The polyphenol metabolites can reach the cells via the bloodstream. By binding to the HSA and the cells, the availability of the phenolic metabolites may be reduced.[52,14] Through the association to proteins, the resulting metabolites or even the sensitive anthocyanin aglycones may be protected from degradation.[22]

2.3.3.6 Cells

Possibly an altered pH value or interactions with the cell membrane can induce the dissolution of the phenol-HSA complex. Whether any phenolic compounds can be taken up into the cells depends on their form, e.g., methylated flavan-3-ols are taken up into cells, whereas glucuronidated flavan-3-ols are not.[35] Binding to the target cells probably occurs via a hydrogen bond between the polar phospholipid head of the cell membrane and the hydroxyl group of the phenol, whereby a higher number of hydroxyl groups on a phenol could provide a stronger bond.[14] It is still unclear whether certain phenols only bind to specific target cells. However, in animal experiments a large number of metabolites in various concentrations have been found in almost all organs.[14] Some mainly non-polar, lipophilic phenolic compounds can even pass the blood-brain barrier.[35]

2.3.3.7 Excretion

The presence of a high level of methylated anthocyanins in rat liver and the low level of these derivatives in plasma suggests that these metabolites may be excreted directly from the liver into the bile, as previously suspected. The final site anthocyanins pass before excretion is the kidney.

Here, metabolism also occurs, with methylation appearing to be the main pathway. However, UGT has also been detected in kidney tissue and glucuronides have been detected accordingly. [53]

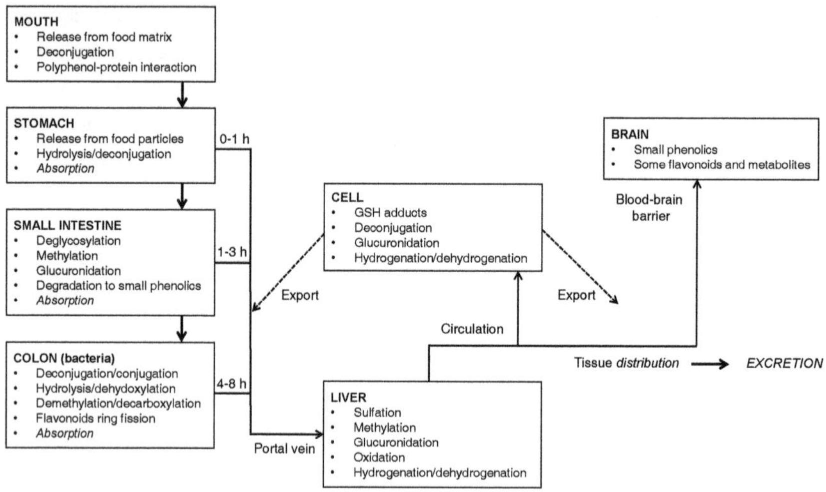

Figure 5: Absorption, distribution, metabolism and excretion of polyphenols[35]

2.3.4 Phase II metabolism

Phase II enzymes, such as UDP-glucuronosyltransferases (UGT) and sulfotransferases (SULT), catalyze the functionalization of activated or nucleophilic compounds, such as anthocyanins, by conjugation with electrophilic molecules such as glucuronic acid or sulfate residues. (**Figure 6**) This allows inactivation of these compounds, which usually become more polar and thus may be more easily excreted via the kidneys and bile. Enzymes like the catechol-*O*-methyltransferases (COMT) are mainly used for the masking of reactive functional groups (e.g. -SH, -OH and -NH).

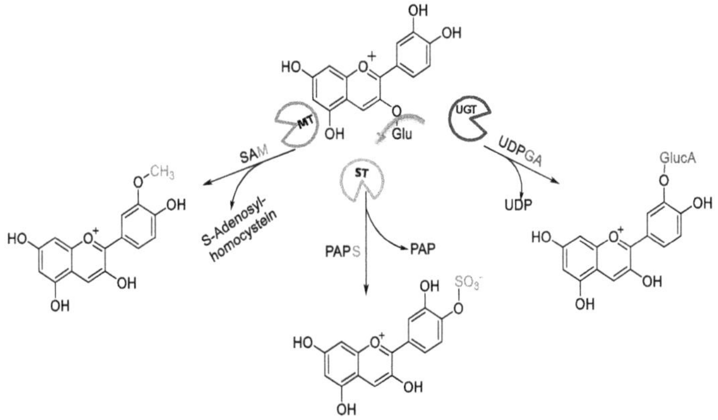

Figure 6: Possible phase II reactions of cyanidin-3-O-glucoside. MT: Methyltransferase, SAM: S-adenosyl-L-methionine, ST: Sulfotransferase, PAPS: 3'-Phosphoadenosine-5'-phosphosulfate, PAP: 3'-Phosphoadenosine-5'-phosphate, UGT: UDP-glucuronosyltransferase, UDPGA: Uridine diphosphate glucuronic acid

2.3.4.1 UDP-glucuronosyltransferases (UGT)

The UDP-glucuronosyltransferases are a family of membrane-bound enzymes that catalyze the transfer of a UDP-activated glucuronic acid (UDPGA, structure shown in **Figure 7**) to a foreign substance.[54] According to Jancova and co-workers, UGTs in humans are divided into four subfamilies.[54] The UGT1 and UGT2 subfamilies are mainly responsible for the glucuronidation of endo- and xenobiotics and thus occupy the most important position in the metabolism. The UGT's are widely distributed and found in many hepatic and extrahepatic tissues, but the liver is the main site of metabolism.[55] A total of 22 different UGT enzymes have been identified in humans, of which the enzymes UGT1A1, 1A3, 1A4, 1A6, 1A9, 2B7 and 2B15 are considered most important for foreign metabolism in humans.[54] In general, the formation of glucuronide conjugates has been shown to be the most common phase II metabolic pathway for water soluble metabolites. This applies to the majority of conjugated metabolites found in urine and bile.[56] Glucuronidation experiments with liver microsomes have shown that strong interspecies differences in the formation of glucuronide metabolites and stereoselectivity of the reaction can occur between humans, monkeys, pigs, dogs and rats.[55,54,57]

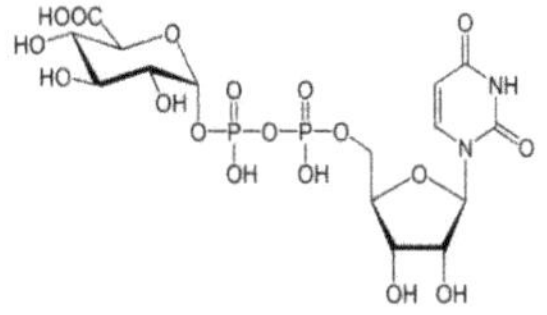

Figure 7: Uridine diphosphate glucuronic acid (UDPGA)

2.3.4.2 Sulfotransferases (SULT)

The sulfotransferases (SULT) catalyze the transfer of a sulfonate group (SO_3^-) via the cofactor 3'-phosphoadenosine-5'-phosphosulfate (PAPS or activated sulfate, structure shown in **Figure 8**) to a functional group (-OH, -SH, $-NH_2$, etc.) of a substance.[58] All human cells are capable of synthesizing PAPS in the cytoplasm in a two-step reaction, using ATP and inorganic sulfate.[59] In most cases, the transfer of the sulfonate group increases the water solubility of the xenobiotic and usually reduces its biological activity.[60] Sulfated foreign substances can then be better excreted in the urine. At a physiological pH value of 7.4, sulfonate conjugates are present in an ionized form, which increases the water solubility and reduces reabsorption due to the reduction in membrane permeability caused by the negative charge. The sulfotransferases can be either cytosolic or membrane-bound at the Golgi apparatus, whereby the cytosolic enzymes predominantly derivatize smaller endogenous and exogenous substances and the membrane-bound sulfotransferases predominantly sulfate hydrophilic substances such as carbohydrates and endogenous proteins, peptides and lipids.[60,61] Inhibition at high substrate concentrations is characteristic for sulfotransferases. The inactivation is probably caused by the simultaneous binding of two foreign molecules at the binding site, which inhibits enzyme activity.[60] Sulfation and glucuronidation of xenobiotics are competing reactions in metabolism, and at high substrate concentrations glucuronidation is preferred due to the lower limitation of UDPGA, whereas PAPS is consumed very quickly in the cell. The sulfation of foreign substances has a lower capacity but is characterized by a high substrate affinity. At lower concentrations, sulfation is therefore more efficient due to the higher binding affinity and allows faster excretion than glucuronidation.[62] Human sulfotransferases can be divided into four different subfamilies: SULT1, SULT2, SULT4 and SULT6.[54] The SULT1 subfamily seems to be the most important subfamily for the sulfation of xenobiotics due to its broad substrate specificity.[54] SULT1A1 is the predominant SULT isoform in the human liver[63], but could also be found in many other tissues, such as brain, breast, intestine, jejunum, lungs, adrenal glands, endometrium, placenta, kidney and platelets[54]. According to Jancova et al., however, there is a lack of information regarding interspecies differences with regard to sulfotransferases.[54] In some species SULTs could be identified that do not resemble any human isoform. This also explains differences in the formation of xenobiotics between humans and animals.

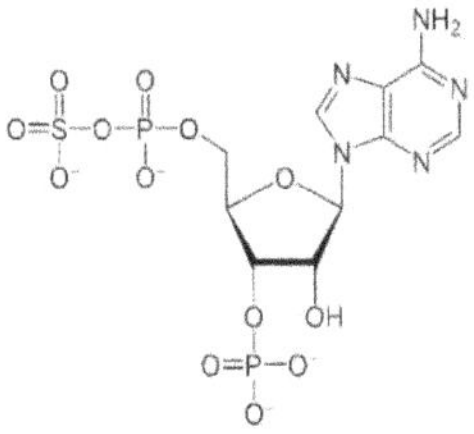

Figure 8: 3'-Phosphoadenosine-5'-phosphosulfate (PAPS)

2.3.4.3 Catechol-*O*-methyltransferase (COMT)

Catechol-*O*-methyltransferases (COMT) can be found in most mammalian tissues, but the highest activities are found in the liver, kidney, and intestinal tract.[64] In humans, two isoforms of COMT are known: one is the membrane-bound form (MB-COMT) and the other is the soluble form (S-COMT), which is found on the cytosolic side of the rough endoplasmic reticulum.[65,54] COMT is an enzyme that catalyzes the transfer of a methyl group of S-adenosyl-L-methionine (SAM, structure shown in **Figure 9**) to the phenolic hydroxyl group of a foreign substance with a catechol structure in the presence of Mg^{2+} ions.[64] The catechol structure is essential for COMT activity. This explains why no methylated derivatives were found for pelargonidin, peonidin and malvidin.[13] Methylation first makes substances less hydrophilic, which can increase their membrane permeability. For effective excretion, however, additional sulfation or glucuronidation is often necessary.[29]

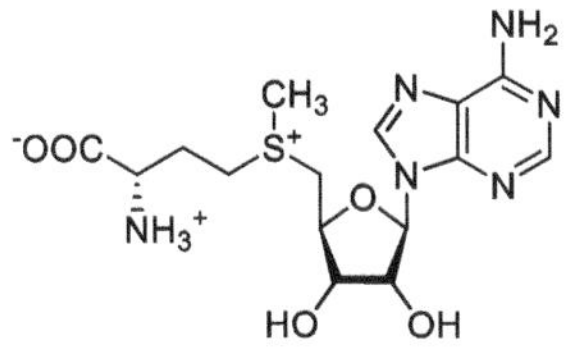

Figure 9: S-adenosyl-L-methionine (SAM)

3 Synthesis of phase II metabolites

It has not yet been established with certainty whether it is actually the anthocyanins themselves or their metabolites that have the effects described in section 2.3. Therefore, it is necessary to accurately identify and quantify the anthocyanin metabolites. For this purpose, reference substances are urgently needed. Because, they are not commercially available, their synthesis is indispensable. For the synthesis of methylated, sulfated and glucuronidated phenolic metabolites many possible ways are conceivable. The enzymatic, chemical, and microbial syntheses of the metabolites are presented below.

The phenolic metabolites could also be extracted directly from plant material, for example, quercetin glucuronides[66] and sulfates[67], or isolated from blood or urine after consumption of food containing phenolic substances. However, especially in the case of anthocyanins, the synthesis is the sole way to obtain them, as their content in physiological fluids is too low and they are not known to occur in plants. Therefore, it is of great importance to develop synthesis strategies so that the phenolic metabolites can be produced in high quantities, cheaply and quickly.

3.1 Enzymatic synthesis

The required metabolites can be produced by enzymes. UDP-glucuronosyltransferases, sulfotransferases and *O*-methyltransferases, whose functions in the body are explained in section 2.3.4, find application for this purpose. These enzymes can be purchased in pure form or obtained from animal or human liver cells.[68,69,63,70] Homogenized liver can be divided into different fractions: S9-Mix, cytosolic and microsomal fraction. The S9 mix represents liver homogenate after removal of cell nuclei and mitochondria. This is achieved by centrifuging homogenized liver at 9000 *g* for 10 minutes and using the supernatant. In this mix all three types of enzymes are contained. Centrifuging this supernatant for one hour at 100000 *g* results in the cytosolic fraction (supernatant), containing COMT and SULT enzymes and the microsomal fraction (pellet) comprising COMT and UGT enzymes.[71] Attempts to sulfate polyphenols with arylsulfotransferase from human intestinal bacteria or plant sulfotransferases have also been successful.[72,73] However, it should be noted that the metabolites produced depend on the origin of the enzymes and the processing and storage of the starting material[63,74,51] The advantages of this method are its stereospecificity and the fact that, unlike most chemical syntheses, it requires only a single step. Disadvantages are the enzyme specificity and a relatively low yields and high prices.

The enzymatic hemisynthesis conducted in the context of this work was performed with porcine liver enzymes and can be found in **chapter 2.**

3.2 Chemical synthesis

The advantage of chemical synthesis over enzymatic synthesis is the higher yield. Besides, the reactants are cheaper than the enzymes and their cofactors. In the case of the present work, one of the starting materials, cyanidin-3-*O*-glucoside could even be obtained from natural sources, namely from blackberry juice. Although it is not possible to exactly replicate the body's metabolism via chemical synthesis, it is possible to isolate the desired derivatives by purification following synthesis. Another way to produce specific derivatives is the introduction of protective groups. These can block certain parts of the molecule so that only the desired hydroxyl groups are free to react and so the required derivative is produced. However, to keep the number of reaction steps as small as possible and since it has not yet been determined exactly at which positions in the molecule the derivatization takes place in the body, this technique should be regarded with less attention. Nevertheless, it is a possibility to produce individual derivatives in a very targeted manner.

3.2.1 Total synthesis

The synthesis of compounds with a pyrylium structure has been known since the early 20th century. In 1922, Pratt and Robinson published the first synthesis specification for pyrylium salts of the anthocyanidin type.[75] The known synthesis pathways of the last decades contain two molecules that form the A and the B rings and lead to the corresponding flavylium ion by cyclization of the C ring (for the designation see **Figure 1**). The two parts are called 'eastern' and 'western' portions of the molecule. The conditions for these synthesis reactions are, however, unfavorable for anthocyanins due to the high pH values. Therefore, there are very few reports about them. Cruz et al.[76] were the first to develop a strategy based on Robinson's acidic aldol condensation to produce anthocyanin metabolite derivatives. First, they succeeded in producing the cyanidin-4'-*O*-methyl-3-glucoside, and later the malvidin-3-*O*-β-glucuronide.[77,76] These reactions sound promising, but they consist of many steps. Each step in a reaction can lead to a reduction in yield, since the conversion rate for the individual steps is rarely 100%. However, the aim should be to achieve a reaction that is as fast as possible with the simplest and least possible steps, which leads to the desired result without many reactants and with the best conversion rate. For these reasons, and because the cyanidin-3-*O*-glucoside could be isolated from blackberry juice, this work has focused mainly on hemisynthesis.

3.2.2 Sulfation

The first attempts to obtain flavonoid sulfates were made by Barron et al. in 1985.[78] They isolated quercetin-3-sulfate and patuletin-3-sulfate from plants. Two years later, they presented the first flavonoid sulfate chemical syntheses with *N,N'*-dicyclohexylcarbodiimides and tetrabutylammonium hydrogen sulfate as reagents[79] and enzymatic syntheses with arylsufotransferases isolated from plants[67]. Methods using sulfamic acid as sulfating reagent were applied earlier but they only provided low yields.[67] However, they were later described as successful in sulfating quercetin.[80] The use of sulfur trioxide adducts, which were not perfectly suited at first[67], was later picked up and optimized to generate quercetin sulfates in dioxane[81]. Since the latter reaction pathway was relatively often described as successful for the sulfation of flavonoids, it was chosen for the sulfation of anthocyanins in this work. To the best of our knowledge, chemical sulfation of anthocyanins has not yet been reported in the literature. A detailed description of the hemisynthesis experiments carried out can be found in chapter 3.

3.2.3 Methylation

In general, the methylation of phenolic compounds by methyl iodide or dimethyl sulfate is well known.[82,83] A less known, but environmentally friendly method is the methylation through dimethyl carbonate with the help of a catalyst.[84] This methods advantage is the use of less toxic reactants than the commonly known and was chosen for the methylation of anthocyanins due to the above-mentioned properties (**Chapter 4**).

3.2.4 Glucuronidation

Glucuronidation of phenolic substances is possible in different ways. The Koenigs-Knorr reaction represents a synthesis of alkyl and aryl *O*-glycosides starting from a glycosyl halide as glycosyl donor and an alcohol or phenol as glycosyl acceptor in the presence of a heavy metal salt (or a Lewis acid) as co-activator (**Figure 10**). Mechanistically, this is a nucleophilic substitution.[85] Among others, glucuronic acid can also be glycosidically bound to organic residues such as flavonoids.[86] Successful examples are the glucuronidation of quercetin[87,88] and of epicatechin[89]. All syntheses use the Koenigs-Knorr reaction to form glucuronides.

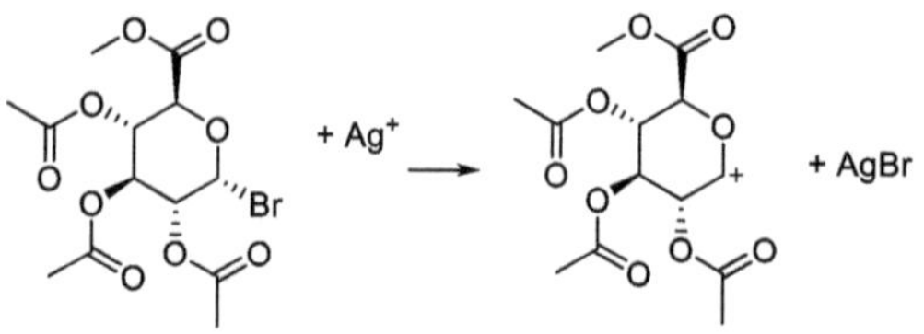

+ Ag^+ ⟶ + AgBr

Figure 10: First part of the Koenigs-Knorr reaction with acetobromo glucuronic acid

It is known that the synthesis of glucuronides is more difficult than the synthesis of the corresponding glucopyranosides, since glucuronyl donors require the highest activation in comparison to the corresponding glucopyranosides in order to cause a reaction.[90] In 1994, Müller et al.[91] published a difficulty scale for the production of various glycosidically bound compounds. The synthesis of glucuronides is described as the most difficult: There is no single powerful method for either glycosylations or glucuronidations. Rather, a special method must often be developed based on the given reactants, taking into account their respective properties.[92] Concluding from the difficulty of glucuronidation and the results of some tests carried out so far, it can be assumed that acetobromoglucuronic acid methyl ester is not suitable for glucuronidation of cyanidin-3-*O*-glucoside under the tested conditions. (unpublished results) As a further outlook the replacement of the glucuronic acid donor should be mentioned. The structurally identical donor, substituted with iodine instead of bromine, should be more reactive since iodine has better leaving group abilities.[93] Instead of glucuronyl halides, the more efficient glucuronyl trichloroacetimidates[94] (**Figure 11**) or even more reactive glucuronyl trifluoroacetimidates could be used[95]. This approach already showed first positive results in preliminary tests.

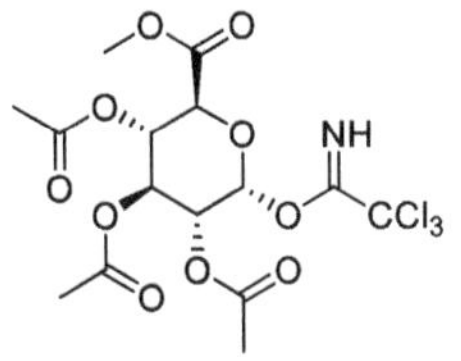

Figure 11: Trichloroacetimidate

Two alternative glucuronidation methods should be mentioned. One is the oxidation of glucose directly to glucuronic acid, while protecting other functional groups by appropriate measures.[96] One argument against this method is that anthocyanins, with their antioxidant properties, are likely to be quite sensitive to an oxidation reaction. Another method is the use of the Koenigs-Knorr reaction for glucuronidation of the split heterocycle followed by aldol condensation to flavonoid.[97] This way has also been successfully applied to the synthesis of cyanidin-7-*O*-glucuronide-3-*O*-glucoside.[17] Malvidin-3-*O*-glucuronide was also synthesized via a condensation reaction.[77]

3.3 Microbial synthesis

Microorganisms are able to form extracellular enzymes, which can lead to glycosylation, deglycosylation, methylation, sulfation or glucuronidation of phenolic compounds.[70] The enzymes glycosyltransferase from *Bacillus cereus*, glucansucrase from *Leuconostoc mesenteroides*, cellulase from *Penicillium decumbrens* or *Aspergillus niger*, and α-amylase from *Bacillus* sp. are able to glycosylate quercetin and catechin, among others.[98–101] Deglycosylation is based on the glycosyl hydrolases β-glucosidase, naringinase, α-rhamnosidase and hesperidinase. These break down the glycosidic bond, resulting in a lipophilic, more unstable aglycone.[102–105] Ring splitting can be performed by enzymes of fungi of the genera *Aspergillus, Penicillium, Rhizopus* or *Monascus*, which convert the phenols into hydroxylated chalcones.[106]

Particularly relevant for the synthesis of phenolic phase II metabolites are methylation, sulfation and glucuronidation. Methylation and glucuronidation were already achieved with quercetin through *Beauvaria bassiana*, a parasitic fungus. For rutin all three conversions were achieved with the filamentous fungus *Cunninghamella echinulata*.[107] The glucuronidation of rutin, quercetin, naringenin and naringin was successful using *Streptomyces* sp..

Glucuronidation by *Streptomyces* sp. is not regiospecific; 3-*O*-β-D-glucuronides, 4-*O*-β-D-glucuronides and 7-*O*-β-D-glucuronides are produced .[49] In addition to the already mentioned *Cunninghamella echinulata*, sulfation was also achieved by *Cunninghamella blakesleeana*, *Streptomyces fulvissimus*, and *Mucor ramannianus*. It produces 4' or 7' sulfates of rutin, 5-hydroxyflavone, kaempferol and hespertin.[106,108,109] The methylation using *Streptomyces griseus* was already successful for (+)-catechin. The catechol-*O*-methyltransferase leads to the formation of 3'-*O*-methyl-(+)-catechin as well as 3',4'-*O*-dimethyl-(+)-catechin. In experiments with quercetin, different compounds were formed which were methylated up to five times.[110–112] Methyltransferases are extremely substrate specific, microorganisms that can methylate catechin are therefore not necessarily able to methylate other polyphenols.[110] Microbial synthesis can already ferment some polyphenols, resulting in a high yield of different, desired metabolites.[49] Microbial synthesis of anthocyanin phase II metabolites has not been reported in the literature yet. Experiments with *Streptomyces tubercidicus* were promising regarding catechin, but cyanidin-3-*O*-glucoside has not been transformed by its enzymes.

4 Analysis and Characterization of the metabolites

The characterization of anthocyanins is difficult due to the wide range of different structures. The balance between colored and colorless isoforms, pH, the presence of copigments and high temperatures are decisive factors for the analysis of anthocyanins in complex matrices.[10] Conjugation at certain sites is described as promoting stability[113], but since there are no sufficient reference materials available yet, little is known about the stability of anthocyanin metabolites. The characterization of the metabolites includes the type and number of conjugated groups and their exact position. For this purpose, spectroscopic methods like UV/Vis or MS are best suited as through conjugation the photometric properties and the mass and fragmentation patterns change (see **chapters 2**, **3** and **4**). As most of the substances of interest occur in complex mixtures, liquid chromatography techniques equipped with a diode array and/or mass spectrometric detectors are a good option to access the spectra of the single substances in a mixture on-line. In this way many useful data can be obtained. NMR is also a particularly useful technique, although it should be noted that larger quantities of substance with higher purities are required.

4.1 UV/Vis spectra

As described in section 2.2, flavonoids and especially anthocyanins have characteristic UV/Vis spectra, which can be changed through various conjugations. The resulting spectra may tentatively be used to identify and characterize flavonoids and their metabolites, as many authors describe shifts through methylation, sulfation or glucuronidation.[114,115] A description of this and the influence of the specific groups on the spectra can be found in **chapters 3** and **4**.

4.2 Mass spectrometry

As most flavonoids and particularly anthocyanins are relatively small polar compounds, atmospheric pressure ionization, especially electrospray ionization, finds application for their analysis through mass spectrometry.[18] Since anthocyanins are already positively charged, positive ionization is useful here. The ionization of the precursor molecule initially produces the $[M]^+$ form. Most molecules then lose the sugar residue through the first fragmentation resulting in the $[M-162]^+$. The following loss is in most cases that of the conjugated group. This is SO_3 ($[M-162-80]^+$) in sulfated, a glucuronic acid ($[M-162-176]^+$) in glucuronidated and CH_3 ($[M-162-14]^+$ or $[M-162-15]^+$) in methylated molecules.[116] When the ionization conditions are optimal, smaller characteristic fragments can be formed through, for example, Retro Diels-Alder cleavage, which still contain the respective group and can be used for identification.[116] Through LC-MS techniques it is possible to determine the number of conjugated groups in the synthesized molecule. In some case the distinction between A, B, or C ring conjugation seems feasible. A distinction between 3' and 4' or 5 and 7 isomers is only possible in combination with other analytical techniques. There are different tandem MS techniques that both have their specific advantages. Ion traps (tandem in time) collect every ion in a sample and are able to fragment these. In this way no metabolite remains undetected, and it is possible to gain an overview of the sample composition. Through the tandem in space measurement with a quadrupole it is possible to search for certain masses and transitions with very high sensitivity. Both techniques found application in this work.

4.2.1 HRMS / IMS

High-resolution mass spectrometry (HRMS) offers a further feature in the analysis of phenolic metabolites. HRMS can distinguish molecules that differ in their *m/z* by only 0.001 u. This accuracy may be reached with a time-of-flight detector, which measures the time differences ions need to pass a flight tube. This is useful, for example, in differentiating between methylated cyanidin glucoside (*m/z* 463.1235) and glucuronidated cyanidin (*m/z* 463.0871).

Coupled with ion mobility spectrometry drift times of the ions can be determined. Drift times are specific times that ions need to pass a drift tube against an inert gas such as nitrogen. They are dependent on the instrument settings, structure like size, charge and shape of the molecule, and the *m/z* but independent of chromatographic conditions like column types, which can compensate, for example, for retention time shifts. From these drift times, independent collision cross sections (CCS values) can be calculated. The CCS values are not matrix dependent and can be measured in the same run with the ion and fragment masses. Through this additional dimension a higher confidence in identification can be reached.[117]

4.3 Purification

The purification of the mixtures of flavonoid metabolites obtained through hemisynthesis is challenging. This critical step is often performed by reversed-phase (semi)preparative chromatography with a gradient elution using an aqueous phase as polar solvent and acetonitrile or methanol as organic solvent.[116,88] As the metabolites often vary only in the position of one functional group, their elution behavior is quite similar, wherefore acid is often added to achieve a better separation. This may lead to the cleavage of some conjugates, which may be enhanced by subsequent evaporation of the resulting fractions.[88] A reasonable pre-purification step may be SPE[116] using an anion exchange material (**Chapter 3**). Thus, ways a separation from the bulk of the starting material is possible. A technique frequently used in chemical synthesis for the purification of reaction products is flash chromatography. When the products have been prepared on a sufficiently large scale, this separation technique is a good option both for purifying the reaction products and for isolating the starting material from natural extracts on a larger scale than preparative HPLC.

5 Aims of the thesis

Secondary plant constituents, in particular the substance class of polyphenols, are associated with protective effects in numerous chronic diseases (section 2.3). The underlying mechanisms are far from being fully understood, among others owing to a lack of knowledge about the bioavailability and metabolism of phenolic compounds.[118] The identification and quantification of metabolites is extremely difficult due to the absence of reference substances. This mainly concerns metabolites of anthocyanins, as these are very sensitive phenolic compounds, little research about their synthesis is published. The chemical and enzymatic synthesis of the possible phenolic metabolites can enable identification and accurate quantification. Thus, phenolic metabolites, especially phase II metabolites should be synthesized, purified, and characterized by mass spectrometry. After optimizing the synthesis and preparation steps with respect to yield and purity, the standard substances should be used in real physiological sample matrices (urine) to characterize previously unknown metabolites. The separation and analysis of the isomeric forms could be performed by coupling UHPLC and tandem mass spectrometry. The results obtained in this way could then be used to generate a database for follow-up projects. In conclusion, a final assessment of the biological effects of especially anthocyanins must also take into account their metabolites and degradation products. In this regard in cell systems, it is important to consider their uptake and possible further metabolism by the cells. The aim of the studies should be to find out whether the anthocyanins themselves or their metabolites are responsible for the reported biological effects. Further investigations are necessary to clarify the exact mechanisms of action of these compounds. The synthesis is not only necessary to identify and quantify the metabolites produced in the body, but also to test their stability, interaction with enzymes and cells and to conduct further physiological studies.

The specific aims of the individual studies (**chapters 2-4**) are

- to reproduce the phase II metabolism of polyphenolic substances in the liver as authentically as possible (**chapter 2**)
- to generate a database for further research (**chapter 2-4**)
- to chemically synthesize sulfated metabolites of cyanidin-3-*O*-glucoside and cyanidin (**chapter 3**)
- to apply the synthesized metabolites as reference substances in real physiological samples (**chapter 3**)
- to chemically synthesize methylated metabolites of cyanidin-3-*O*-glucoside in an environmentally safe and non-toxic way (**chapter 4**)

6 References

(1) Kuhnert, N. Polyphenole: Vielseitige Pflanzeninhaltsstoffe. *Chem. unserer Zeit*, **2013**, *47*, 80–91.

(2) Scalbert, A.; Williamson, G. Dietary intake and bioavailability of polyphenols. *J. Nutr.* **2000**, *130*, 2073S-2085S.

(3) He, J.; Giusti, M. M. Anthocyanins: natural colorants with health-promoting properties. *Annu. Rev. Food Sci. Technol.* **2010**, *1*, 163–187.

(4) Kytridis, V.-P.; Manetas, Y. Mesophyll versus epidermal anthocyanins as potential in vivo antioxidants: evidence linking the putative antioxidant role to the proximity of oxy-radical source. *J. Exp. Bot.* **2006**, *57*, 2203–2210.

(5) Lattanzio, V.; Kroon, P. A.; Quideau, S.; Treutter, D. Plant phenolics– Secondary metabolites with diverse functions. In *Recent advances in polyphenol research;* Daayf, F.; Lattanzio, V., Eds.; Wiley-Blackwell: Oxford, UK, **2008**, 1–35.

(6) Oancea, S.; Anthocyanins, from biosynthesis in plants to human health benefits. *Acta Univ. Cibiniensis, Ser. E: Food Technol.* **2011**, *15*.

(7) Schultz, J. C.; Hunter, M. D.; Appel, H. M. Antimicrobial activity of polyphenols mediates plant-herbivore interactions. In *Plant Polyphenols;* Hemingway, R. W.; Laks, P. E., Eds.; Springer US: Boston, MA, **1992**, 621–637.

(8) Clausen, T. P.; Reichardt, P. B.; Bryant, J. P.; Provenza, F. Condensed tannins in plant defense: A perspective on classical theories. In *Plant Polyphenols;* Hemingway, R. W.; Laks, P. E., Eds.; Springer US: Boston, MA, **1992**, 639–651.

(9) Singla, R. K.; Dubey, A. K.; Garg, A.; Sharma, R. K.; Fiorino, M.; Ameen, S. M.; Haddad, M. A.; Al-Hiary, M. Natural polyphenols: Chemical classification, definition of classes, subcategories, and structures. *J. AOAC Int.* **2019**, *102*, 1397–1400.

(10) Clifford, M. N. Anthocyanins - nature, occurrence and dietary burden. *J. Sci. Food Agric.* **2000**, *80*, 1063–1072.

(11) Hollman, P. C.H. Absorption, bioavailability, and metabolism of flavonoids. *Pharm. Biol.* **2004**, *42*, 74–83.

(12) Kraus, M. *Synthese von 14C-markierten Anthocyanidinen und Studien zur intestinalen Verfügbarkeit von Anthocyanen aus Heidelbeeren (Vaccinium myrtillus L.),* Dissertation, Würzburg, **2006**.

(13) Pojer, E.; Mattivi, F.; Johnson, D.; Stockley, C. S. The case for anthocyanin consumption to promote human health: A review. *Compr. Rev. Food Sci. Food Saf.* **2013**, *12*, 483–508.

(14) Manach, C.; Scalbert, A.; Morand, C.; Rémésy, C.; Jiménez, L. Polyphenols: food sources and bioavailability. *Am. J. Clin. Nutr.* **2004**, *79*, 727–747.

(15) Ververidis, F.; Trantas, E.; Douglas, C.; Vollmer, G.; Kretzschmar, G.; Panopoulos, N. Biotechnology of flavonoids and other phenylpropanoid-derived natural products. Part I: Chemical diversity, impacts on plant biology and human health. *Biotechnol. J.* **2007**, *2,* 1214–1234.

(16) Mazza, G.; Cacace, J. E.; Kay, C. D. Methods of analysis for anthocyanins in plants and biological fluids. *J. AOAC Int.* **2004**, *87,* 129–145.

(17) Cruz, L.; Basílio, N.; Mateus, N.; Pina, F.; Freitas, V. de Characterization of kinetic and thermodynamic parameters of cyanidin-3-glucoside methyl and glucuronyl metabolite conjugates. *J. Phys. Chem. B* **2015**, *119,* 2010–2018.

(18) Castañeda-Ovando, A.; Pacheco-Hernández, M. d. L.; Páez-Hernández, M. E.; Rodríguez, J. A.; Galán-Vidal, C. A. Chemical studies of anthocyanins: A review. *Food Chem.* **2009**, *113,* 859–871.

(19) Giusti, M. M.; Wrolstad, R. E. Characterization and measurement of anthocyanins by UV-visible spectroscopy. *Curr. Protoc. Food Anal. Chem.* **2001**, *00,* F1.2.1-F1.2.13.

(20) Giusti, M. M.; Wrolstad, R. E. Characterization of red radish anthocyanins. *J. Food Sci.* **1996**, *61,* 322–326.

(21) Giusti, M. M.; Rodríguez-Saona, L. E.; Wrolstad, R. E. Molar absorptivity and color characteristics of acylated and non-acylated pelargonidin-based anthocyanins. *J. Agric. Food Chem.* **1999**, *47,* 4631–4637.

(22) Hribar, U.; Ulrih, N. P. The metabolism of anthocyanins. *Curr. Drug Metab.* **2014**, *15,* 3–13.

(23) Ghosh, D.; Konishi, T. Anthocyanins and anthocyanin-rich extracts: role in diabetes and eye function. *Asia Pac. J. Clin. Nutr.* **2007**, *16,* 200–208.

(24) Borkowski, T.; Szymusiak, H.; Gliszczyńska-Rwigło, A.; Rietjens, I. M. C. M.; Tyrakowska, B. Radical scavenging capacity of wine anthocyanins is strongly pH-dependent. *J. Agric. Food Chem.* **2005**, *53,* 5526–5534.

(25) Toufektsian, M.-C.; Lorgeril, M. de; Nagy, N.; Salen, P.; Donati, M. B.; Giordano, L.; Mock, H.-P.; Peterek, S.; Matros, A.; Petroni, K.; Pilu, R.; Rotilio, D.; Tonelli, C.; Leiris, J. de; Boucher, F.; Martin, C. Chronic dietary intake of plant-derived anthocyanins protects the rat heart against ischemia-reperfusion injury. *J. Nutr.* **2008**, *138,* 747–752.

(26) Rice-Evans, C. A.; Miller, N. J.; Paganga, G. Structure-antioxidant activity relationships of flavonoids and phenolic acids. *Free Rad. Biol. Med.* **1996**, *20,* 933–956.

(27) Nijveldt, R. J.; van Nood, E.; van Hoorn, D. E.; Boelens, P. G.; van Norren, K.; van Leeuwen, P. A. Flavonoids: a review of probable mechanisms of action and potential applications. *Am. J. Clin. Nutr.* **2001**, *74,* 418–425.

(28) Liu, W.; Guo, R. Interaction between morin and sodium dodecyl sulfate (SDS) micelles. *J. Agric. Food Chem.* **2005**, *53,* 2890–2896.

(29) Cassidy, A.; Minihane, A.-M. The role of metabolism (and the microbiome) in defining the clinical efficacy of dietary flavonoids. *Am. J. Clin. Nutr.* **2017**, *105,* 10–22.

(30) Ichiyanagi, T.; Shida, Y.; Rahman, M. M.; Hatano, Y.; Konishi, T. Bioavailability and tissue distribution of anthocyanins in bilberry (*Vaccinium myrtillus* L.) extract in rats. *J. Agric. Food Chem.* **2006**, *54,* 6578–6587.

(31) Passamonti, S.; Vrhovsek, U.; Mattivi, F. The interaction of anthocyanins with bilitranslocase. *Biochem. Biophys. Res. Commun.* **2002**, *296,* 631–636.

(32) Mülleder, U.; Murkovic, M.; Pfannhauser, W. Urinary excretion of cyanidin glycosides. *J. Biochem. Biophys. Methods* **2002**, *53,* 61–66.

(33) Manach, C.; Williamson, G.; Morand, C.; Scalbert, A.; Rémésy, C. Bioavailability and bioefficacy of polyphenols in humans. I. Review of 97 bioavailability studies. *Am. J. Clin. Nutr.* **2005**, *81,* 230S-242S.

(34) McGhie, T. K.; Walton, M. C. The bioavailability and absorption of anthocyanins: towards a better understanding. *Mol. Nutr. Food Res.* **2007**, *51,* 702–713.

(35) Corona, G.; Vauzour, D.; Amini, A.; Spencer, J. P.E. The impact of gastrointestinal modifications, blood-brain barrier ransport, and intracellular metabolism on polyphenol bioavailability: An overview. In *Polyphenols in human health and disease;* R. R. Watson, V. R., Preedy, S. Zibadi, Ed.; Academic Press: London, **2014**, 591–604.

(36) Terahara, N. Flavonoids in foods: A review. *Nat. Prod. Commun.* **2015**, *10,* 1934578X1501000.

(37) Medina-Remón, A.; Tresserra-Rimbau, A.; Valderas-Martinez, P.; Estruch, R.; Lamuela-Raventos, R. M. Polyphenol consumption and blood pressure. In *Polyphenols in human health and disease;* R. R. Watson, V. R., Preedy, S. Zibadi, Ed.; Academic Press: London, **2014**, 971–987.

(38) Déprez, S.; Brezillon, C.; Rabot, S.; Philippe, C.; Mila, I.; Lapierre, C.; Scalbert, A. Polymeric proanthocyanidins are catabolized by human colonic microflora into low-molecular-weight phenolic acids. *J. Nutr.* **2000**, *130,* 2733–2738.

(39) Aura, A.-M.; O'Leary, K. A.; Williamson, G.; Ojala, M.; Bailey, M.; Puupponen-Pimiä, R.; Nuutila, A. M.; Oksman-Caldentey, K.-M.; Poutanen, K. Quercetin derivatives are deconjugated and converted to hydroxyphenylacetic acids but not methylated by human fecal flora in vitro. *J. Agric. Food Chem.* **2002**, *50,* 1725–1730.

(40) Barron, D. Recent advances in the chemical synthesis and biological activity of phenolic metabolites. In *Recent advances in polyphenol research;* Daayf, F.; Lattanzio, V., Eds.; Wiley-Blackwell: Oxford, UK, **2008**, 317–358.

(41) Dangles, O.; Dufour, C. Flavonoid–protein binding processes and their potential impact on human health. In *Recent advances in polyphenol research;* Daayf, F.; Lattanzio, V., Eds.; Wiley-Blackwell: Oxford, UK, **2008**, 67–87.

(42) Walle, T. Absorption and metabolism of flavonoids. *Free Radicals Biol. Med.* **2004**, *36,* 829–837.

(43) Kroon, P. A.; Clifford, M. N.; Crozier, A.; Day, A. J.; Donovan, J. L.; Manach, C.; Williamson, G. How should we assess the effects of exposure to dietary polyphenols in vitro? *Am. J. Clin. Nutr.* **2004**, *80,* 15–21.

(44) Crespy, V.; Morand, C.; Manach, C.; Besson, C.; Demigne, C.; Remesy, C. Part of quercetin absorbed in the small intestine is conjugated and further secreted in the intestinal lumen. *Am. J. Physiol.* **1999**, *277,* G120-6.

(45) Kuhnle, G.; Spencer, J. P.; Schroeter, H.; Shenoy, B.; Debnam, E. S.; Srai, S. K.; Rice-Evans, C.; Hahn, U. Epicatechin and catechin are *O*-methylated and glucuronidated in the small intestine. *Biochem. Biophys. Res. Commun.* **2000**, *277,* 507–512.

(46) van der Woude, H.; Boersma, M. G.; Vervoort, J.; Rietjens, I. M. C. M. Identification of 14 quercetin phase II mono- and mixed conjugates and their formation by rat and human phase II in vitro model systems. *Chem. Res. Toxicol.* **2004**, *17,* 1520–1530.

(47) Fleschhut, J.; Kratzer, F.; Rechkemmer, G.; Kulling, S. E. Stability and biotransformation of various dietary anthocyanins in vitro. *Eur. J. Nutr.* **2006,** 7–18.

(48) Nassar, A. F. Approaches to performing metabolite elucidation: One key to success in drug discovery and development. In *Pharmaceutical Sciences Encyclopedia;* Gad, S. C., Ed.; John Wiley & Sons, Inc: Hoboken, NJ, USA, **2010**, 1-23.

(49) Marvalin, C.; Azerad, R. Microbial glucuronidation of polyphenols. *J. Mol. Catal. B: Enzym.* **2011**, *73,* 43–52.

(50) Evans, T. J. Toxicokinetics and toxicodynamics. In *Small Animal Toxicology;* M. E. Peterson, P. A. Talcott, Ed.; Elsevier: Oxford, UK **2006**, 13–19.

(51) Santos-Buelga, C.; González-Manzano, S.; Dueñas, M.; González-Paramás, A. M. Analysis and characterisation of flavonoid phase II metabolites. In *Recent advances in polyphenol research;* Cheynier, V.; Sarni-Manchado, P.; Quideau, S., Eds.; Wiley-Blackwell: Oxford, UK, **2012**, 249–286.

(52) Dangles, O.; Dufour, C.; Bret, S. Flavonol–serum albumin complexation. Two-electron oxidation of flavonols and their complexes with serum albumin. *J. Chem. Soc., Perkin Trans. 2* **1999,** 737–744.

(53) Talavéra, S.; Felgines, C.; Texier, O.; Besson, C.; Gil-Izquierdo, A.; Lamaison, J.-L.; Rémésy, C. Anthocyanin metabolism in rats and their distribution to digestive area, kidney, and brain. *J. Agric. Food Chem.* **2005**, *53,* 3902–3908.

(54) Jancova, P.; Anzenbacher, P.; Anzenbacherova, E. Phase II drug metabolizing enzymes. *Biomed. Pap. Med. Fac. Univ. Palacky Olomouc Czech Repub.* **2010**, *154,* 103–116.

(55) Kaji, H.; Kume, T. Characterization of afloqualone N-glucuronidation: species differences and identification of human UDP-glucuronosyltransferase isoform(s). *Drug Metab. Dispos.* **2005**, *33,* 60–67.

(56) Ferrars, R. de. *The metabolic fate and bioactivity of anthocyanins in humans,* doctoral dissertation, University of East Anglia, **2014**.

(57) Matal, J.; Jancova, P.; Siller, M.; Masek, V.; Anzenbacherova, E.; Anzenbacher, P. Interspecies comparison of the glucuronidation processes in the man, monkey, pig, dog and rat. *Neuroendocrinol. Lett.* **2008**, *29,* 738–743.

(58) Nowell, S.; Falany, C. N. Pharmacogenetics of human cytosolic sulfotransferases. *Oncogene* **2006**, *25,* 1673–1678.

(59) Klaassen, C. D.; Boles, J. W. Sulfation and sulfotransferases 5: the importance of 3'-phosphoadenosine 5'-phosphosulfate (PAPS) in the regulation of sulfation. *FASEB J.* **1997**, *11,* 404–418.

(60) Gamage, N.; Barnett, A.; Hempel, N.; Duggleby, R. G.; Windmill, K. F.; Martin, J. L.; McManus, M. E. Human sulfotransferases and their role in chemical metabolism. *Toxicol. Sci.* **2006**, *90,* 5–22.

(61) Paul, P.; Suwan, J.; Liu, J.; Dordick, J. S.; Linhardt, R. J. Recent advances in sulfotransferase enzyme activity assays. *Anal. Bioanal. Chem.* **2012**, *403,* 1491–1500.

(62) Berg, K. *In vitro Untersuchungen zur Genotoxizität der Alkenylbenzene a-,b und g-Asaron sowie ausgewählter oxidativer Metaboliten,* Dissertation, Kaiserslautern, **2016**.

(63) Vaidyanathan, J. B.; Walle, T. Glucuronidation and sulfation of the tea flavonoid (−)-epicatechin by the human and rat enzymes. *Drug Metab. Dispos.* **2002**, *30,* 897–903.

(64) Casal, E.; Palomo, L.; Cabrera, D.; Falcon-Perez, J. M. A novel sensitive method to measure catechol-*O*-methyltransferase activity unravels the presence of this activity in extracellular vesicles released by rat hepatocytes. *Front. Pharmacol.* **2016**, *7,* 501.

(65) Ulmanen, I.; Peränen, J.; Tenhunen, J.; Tilgmann, C.; Karhunen, T.; Panula, P.; Bernasconi, L.; Aubry, J. P.; Lundström, K. Expression and intracellular localization of catechol *O*-methyltransferase in transfected mammalian cells. *Eur. J. Biochem.* **1997**, *243,* 452–459.

(66) Parejo, I.; Viladomat, F.; Bastida, J.; Schmeda-Hirschmann, G.; Burillo, J.; Codina, C. Bioguided isolation and identification of the nonvolatile antioxidant compounds from fennel (*Foeniculum vulgare* Mill.) waste. *J. Agric. Food Chem.* **2004**, *52,* 1890–1897.

(67) Barron, D.; Varin, L.; Ibrahim, R. K.; Harborne, J. B.; Williams, C. A. Sulphated flavonoids – an update. *Phytochemistry* **1988**, *27,* 2375–2395.

(68) Day, A. J.; Bao, Y.; Morgan, M. R.A.; Williamson, G. Conjugation position of quercetin glucuronides and effect on biological activity. *Free Radicals Biol. Med.* **2000**, *29,* 1234–1243.

(69) Boersma, M. G.; van der Woude, H.; Bogaards, J.; Boeren, S.; Vervoort, J.; Cnubben, N. H. P.; van Iersel, M. L. P. S.; van Bladeren, P. J.; Rietjens, I. M. C. M. Regioselectivity of phase II metabolism of luteolin and quercetin by UDP-glucuronosyl transferases. *Chem. Res. Toxicol.* **2002**, *15,* 662–670.

(70) Huynh, N. T.; van Camp, J.; Smagghe, G.; Raes, K. Improved release and metabolism of flavonoids by steered fermentation processes: a review. *Int. J. Mol. Sci.* **2014**, *15,* 19369–19388.

(71) Rasmussen, M. K.; Ekstrand, B.; Zamaratskaia, G. Comparison of cytochrome P450 concentrations and metabolic activities in porcine hepatic microsomes prepared with two different methods. *Toxicol. In Vitro* **2011,** 343–346.

(72) Varin, L.; Barron, D.; Ibrahim, R. K. Enzymatic assay for flavonoid sulfotransferase. *Anal. Biochem.* **1987**, *161,* 176–180.

(73) Koizumi, M.; Shimizu, M.; Kobashi, K. Enzymatic sulfation of quercetin by arylsulfotransferase from a human intestinal bacterium. *Chem. Pharm. Bull.* **1990**, *38,* 794–796.

(74) Dueñas, M.; Mingo-Chornet, H.; Pérez-Alonso, J. J.; Di Paola-Naranjo, R.; González-Paramás, A. M.; Santos-Buelga, C. Preparation of quercetin glucuronides and characterization by HPLC–DAD–ESI/MS. *Eur. Food Res. Technol.* **2008**, *227,* 1069–1076.

(75) Pratt, D. D.; Robinson, R. A synthesis of pyrylium salts of anthocyanidin type. *J. Chem. Soc., Abstr.* **1922,** 1577–1585.

(76) Cruz, L.; Mateus, N.; Freitas, V. de First chemical synthesis report of an anthocyanin metabolite with in vivo occurrence: cyanidin-4′-*O*-methyl-3-glucoside. *Tetrahedron Lett.* **2013**, *54,* 2865–2869.

(77) Cruz, L.; Fernandes, I.; Évora, A.; Freitas, V. de; Mateus, N. Synthesis of the main red wine anthocyanin metabolite: Malvidin-3-*O*-β-glucuronide. *Synlett* **2017**, *28,* 593–596.

(78) Barron, D.; Colebrook, L. D.; Ibrahim, R. K. An equimolar mixture of quercetin 3-sulphate and patuletin 3-sulphate from Flaveria chloraefolia. *Phytochemistry* **1986**, *25,* 1719–1721.

(79) Barron, D.; Ibrahim, R. K. Synthesis of flavonoid sulfates: 1. stepwise sulfation of positions 3, 7, and 4 using *N,N'*-dicyclohexylcarbodiimide and tetrabutylammonium hydrogen sulfate. *Tetrahedron* **1987**, *43,* 5197–5202.

(80) Day, A. J.; Mellon, F.; Barron, D.; Sarrazin, G.; Morgan, M. R.; Williamson, G. Human metabolism of dietary flavonoids: identification of plasma metabolites of quercetin. *Free Rad. Res.* **2001**, *35,* 941–952.

(81) Jones, D. J. L.; Jukes-Jones, R.; Verschoyle, R. D.; Farmer, P. B.; Gescher, A. A synthetic approach to the generation of quercetin sulfates and the detection of quercetin 3'-*O*-sulfate as a urinary metabolite in the rat. *Bioorg. Med. Chem.* **2005**, *13,* 6727–6731.

(82) Bouktaib, M.; Lebrun, S.; Atmani, A.; Rolando, C. Hemisynthesis of all the *O*-monomethylated analogues of quercetin including the major metabolites, through selective protection of phenolic functions. *Tetrahedron* **2002**, *58,* 10001–10009.

(83) Cren-Olivé, C.; Lebrun, S.; Rolando, C. An efficient synthesis of the four mono methylated isomers of (+)-catechin including the major metabolites and of some dimethylated and trimethylated analogues through selective protection of the catechol ring. *J. Chem. Soc., Perkin Trans. 1* **2002,** 821–830.

(84) Shieh, W. C.; Dell, S.; Repic, O. 1,8-Diazabicyclo5.4.0undec-7-ene (DBU) and microwave-accelerated green chemistry in methylation of phenols, indoles, and benzimidazoles with dimethyl carbonate. *Org. Lett.* **2001**, *3,* 4279–4281.

(85) Kürti, L.; Czakó, B. *Strategic applications of named reactions in organic synthesis. Background and detailed mechanisms; 250 named reactions;* Elsevier Acad. Press: Amsterdam, **2009**.

(86) Docampo, M.; Olubu, A.; Wang, X.; Pasinetti, G.; Dixon, R. A. Glucuronidated flavonoids in neurological protection: Structural analysis and approaches for chemical and biological synthesis. *J. Agric. Food Chem.* **2017**, *65,* 7607–7623.

(87) Wagner, H. Isolierung von Kämpferol-3-ß-D-glucuronid aus *Euphorbia esula* L. und seine Synthese. *Chem. Ber.* **1970**, *103,* 3678.

(88) Needs, P. W.; Kroon, P. A. Convenient syntheses of metabolically important quercetin glucuronides and sulfates. *Tetrahedron* **2006**, *62,* 6862–6868.

(89) Zhang, M.; Jagdmann, G. E.; van Zandt, M.; Sheeler, R.; Beckett, P.; Schroeter, H. Chemical synthesis and characterization of epicatechin glucuronides and sulfates: bioanalytical standards for epicatechin metabolite identification. *J. Nat. Prod.* **2013**, *76,* 157–169.

(90) Fan, J.; Brown, S. M.; Tu, Z.; Kharasch, E. D. Chemical and enzyme-assisted syntheses of norbuprenorphine-3-β-D-glucuronide. *Bioconjugate Chem.* **2011**, *22,* 752–758.

(91) Müller, T.; Schneider, R.; Schmidt, R. R. Utility of glycosyl phosphites as glycosyl donors- fructofuranosyl and 2-deoxyhexopyranosyl phosphites in glycoside bond formation. *Tetrahedron Lett.* **1994**, *35,* 4763–4766.

(92) Toshima, K., Tatsuta, K.; Recent progress in O-glycosylation methods and its application to natural products synthesis. *Chem. Rev.* **1993**, *93,* 1503–1531.

(93) Paulsen, H. Synthesis of complex oligosaccharide chains of glycoproteins. *Chem. Soc. Rev.* **1984**, *13,* 15.

(94) Pearson, A. G.; Kiefel, M. J.; Ferro, V.; Itzstein, M. von Towards the synthesis of aryl glucuronides as potential heparanase probes. An interesting outcome in the glycosidation of glucuronic acid with 4-hydroxycinnamic acid. *Carbohydr. Res.* **2005**, *340,* 2077–2085.

(95) Al-Maharik, N.; Botting, N. P. A facile synthesis of isoflavone 7-O-glucuronides. *Tetrahedron Lett.* **2006**, *47,* 8703–8706.

(96) Kajjout, M.; Rolando, C. Regiospecific synthesis of quercetin O-β-D-glucosylated and O-β-D-glucuronidated isomers. *Tetrahedron* **2011**, *67,* 4731–4741.

(97) Luis, J. G. Synthesis of danielone (α-hydroxyacetosyringone). *J. Chem. Res., Synop.* **1999**, *999,* 220.

(98) Rao, K. V.; Weisner, N. T. Microbial transformation of quercetin by *Bacillus cereus*. *Appl. Environ. Microbiol.* **1981**, *42,* 450–452.

(99) Gao, C.; Mayon, P.; MacManus, D. A.; Vulfson, E. N. Novel enzymatic approach to the synthesis of flavonoid glycosides and their esters. *Biotechnol. Bioeng.* **2000**, *71,* 235–243.

(100) Bertrand, A.; Morel, S.; Lefoulon, F.; Rolland, Y.; Monsan, P.; Remaud-Simeon, M. *Leuconostoc mesenteroides* glucansucrase synthesis of flavonoid glucosides by acceptor reactions in aqueous-organic solvents. *Carbohydr. Res.* **2006**, *341,* 855–863.

(101) Chen, S.; Xing, X.-H.; Huang, J.-J.; Xu, M.-S. Enzyme-assisted extraction of flavonoids from *Ginkgo biloba* leaves: improvement effect of flavonol transglycosylation catalyzed by *Penicillium decumbens* cellulase. *Enzyme Microb. Technol.* **2011**, *48,* 100–105.

(102) Zheng, Z.; Shetty, K. Solid-state bioconversion of phenolics from cranberry pomace and role of *Lentinus edodes* beta-glucosidase. *J. Agric. Food Chem.* **2000**, *48,* 895–900.

(103) Vattem, D.A.; Lin, Y.-T.; Labbe, R.G.; Shetty, K. Phenolic antioxidant mobilization in cranberry pomace by solid-state bioprocessing using food grade fungus *Lentinus edodes* and effect on antimicrobial activity against select food borne pathogens. *Inno. Food Sci. Emerging Technol.* **2004**, *5,* 81–91.

(104) Liu, J.-Y.; Yu, H.-S.; Feng, B.; Kang, L.-P.; PANG, X.; Xiong, C.-Q.; Zhao, Y.; Li, C.-M.; Zhang, Y.; Ma, B.-P. Selective hydrolysis of flavonoid glycosides by *Curvularia lunata*. *Chin. J. Nat. Med.* **2014**, *11,* 684–689.

(105) Lin, S.; Zhu, Q.; Wen, L.; Yang, B.; Jiang, G.; Gao, H.; Chen, F.; Jiang, Y. Production of quercetin, kaempferol and their glycosidic derivatives from the aqueous-organic extracted residue of litchi pericarp with *Aspergillus awamori*. *Food Chem.* **2014**, *145,* 220–227.

(106) Das, S.; Rosazza, J. P. N. Microbial and enzymatic transformations of flavonoids. *J. Nat. Prod.* **2006**, *69,* 499–508.

(107) Araújo, K. C. F.; M B Costa, E. M. de; Pazini, F.; Valadares, M. C.; Oliveira, V. de Bioconversion of quercetin and rutin and the cytotoxicity activities of the transformed products. *Food Chem. Toxicol.* **2013**, *51,* 93–96.

(108) Ibrahim, A.; Khalifa, S. I.; Khafagi, I.; Youssef, D. T.; Khan, S.; Mesbah, M.; Khan, I. Microbial metabolism of biologically active secondary metabolites from *Nerium oleander* L. *Chem. Pharm. Bull.* **2008**, *56,* 1253–1258.

(109) Herath, W.; Khan, I. A. Microbial metabolism. Part 13. Metabolites of hesperetin. *Bioorg. Med. Chem. Lett.* **2011**, *21,* 5784–5786.

(110) Dhar, K.; Rosazza, J. P. Purification and characterization of *Streptomyces griseus* catechol *O*-methyltransferase. *Appl. Environ. Microbiol.* **2000**, *66,* 4877–4882.

(111) Hosny, M.; Dhar, K.; Rosazza, J. P. Hydroxylations and methylations of quercetin, fisetin, and catechin by *Streptomyces griseus*. *J. Nat. Prod.* **2001**, *64,* 462–465.

(112) Moridani, M. Y.; Scobie, H.; Salehi, P.; O'Brien, P. J. Catechin metabolism: glutathione conjugate formation catalyzed by tyrosinase, peroxidase, and cytochrome p450. *Chem. Res. Toxicol.* **2001**, *14,* 841–848.

(113) Tomaz, I.; Šikuten, I.; Preiner, D.; Andabaka, Ž.; Huzanić, N.; Lesković, M.; Karoglan Kontić, J.; Ašperger, D. Stability of polyphenolic extracts from red grape skins after thermal treatments. *Chem. Pap.* **2019**, *73,* 195–203.

(114) Anouar, E. H.; Gierschner, J.; Duroux, J.-L.; Trouillas, P. UV/Visible spectra of natural polyphenols: A time-dependent density functional theory study. *Food Chem.* **2012**, *131,* 79–89.

(115) Dueñas, M.; González-Manzano, S.; Surco-Laos, F.; González-Paramas, A.; Santos-Buelga, C. Characterization of sulfated quercetin and epicatechin metabolites. *J. Agric. Food Chem.* **2012**, *60,* 3592–3598.

(116) González-Manzano, S.; González-Paramás, A.; Santos-Buelga, C.; Dueñas, M. Preparation and characterization of catechin sulfates, glucuronides, and methylethers with metabolic interest. *J. Agric. Food Chem.* **2009**, *57,* 1231–1238.

(117) Stow, S. M.; Causon, T. J.; Zheng, X.; Kurulugama, R. T.; Mairinger, T.; May, J. C.; Rennie, E. E.; Baker, E. S.; Smith, R. D.; McLean, J. A.; Hann, S.; Fjeldsted, J. C. An interlaboratory evaluation of drift tube ion mobility-mass spectrometry collision cross section measurements. *Anal. Chem.* **2017**, *89,* 9048–9055.

(118) Hollman, P. C. H. Unravelling of the health effects of polyphenols is a complex puzzle complicated by metabolism. *Arch. Biochem. Biophys.* **2014,** 100–105.

(119) Miller, E. R.; Ullrey, D. E. The pig as a model for human nutrition. *Annu. Rev. Nutr.* **1987**, *7,* 361–382.

Chapter 2

Hemisynthesis of Anthocyanin Phase II Metabolites by Porcine Liver Enzymes

The aim of this work was to obtain phase II metabolites of cyanidin-3-*O*-glucoside and its aglycone using porcine liver enzymes. For this purpose, anthocyanins extracted from blackberry concentrate and containing mostly cyanidin-3-*O*-glucoside were incubated with the S9, microsomal, and cytosolic fractions of porcine liver. The reactions were targeted to the direction of the respective phase II transformation by the addition of activated cofactors. LC-MS^n and LC-IMS-QTOF-MS analyses showed that one methylated, three glucuronidated and three sulfated metabolites of cyanidin-3-*O*-glucoside were generated. The aglycone, cyanidin, was sulfated and glucuronidated by the liver enzymes. In addition, both were glucuronidated and methylated simultaneously. The detected compounds and the generated data like exact masses, mass spectra, and CCS values may serve as a basis in the search for metabolites formed in vivo. As their effects are largely unexplored, the described synthesis may contribute to a better understanding of the metabolism of anthocyanins.

Keywords: phase II metabolites, cyanidin, anthocyanins, porcine liver, in vitro, hemisynthesis, polyphenols, ion mobility spectrometry, CCS value

This chapter has been published:

Schmitt, S., Tratzka, S., Schieber, A., Passon, M. Hemisynthesis of Anthocyanin Phase II Metabolites by Porcine Liver Enzymes. *Journal of Agricultural and Food Chemistry* **2019**, 67 (22), 6177-6189.

1 Introduction

Anthocyanins are flavonoids which are consumed in large portions with the intake of juices and fruits. They are polyphenolic secondary plant compounds and responsible for the diverse color of bilberries, blackberries, strawberries, and others.[1] In plants, anthocyanins are present as their glycosides, which are more stable and show a higher solubility in water.[1] The most widespread anthocyanins are the 3-*O*-glycosides and the 3,5-*O*-diglycosides.[2] Numerous studies have linked the intake of anthocyanins to health-promoting effects,[3] such as antioxidant, anti-inflammatory, hypoallergenic, and anticarcinogenic activities, whereas many of the protective properties are attributed to their high antioxidant capacity in vitro.[4] Due to these effects, anthocyanins are increasingly gaining importance, especially in functional food or food supplements. However, there is still a need for further research concerning their absorption, distribution, metabolism, and excretion, which depend on many factors. Anthocyanins are rapidly absorbed in the gastrointestinal tract, but it is assumed that their bioactivity is caused by their metabolites rather than their original structure. In the blood, they can be found as unmodified, methylated, glucuronidated, and/or sulfated forms.[5] However, little is known about the biological activity of the phase II metabolites. The lack of commercially available reference substances or compounds isolated from plants makes synthesis indispensable. Especially the poor commercial availability and the high prices for the phase II metabolites of anthocyanins pose a significant hindrance in the identification and quantification. To quantify the metabolites, alternative, easily obtainable substances such as aglycones are often used.[6] Due to the different ionizability of the aglycone in mass spectrometric analysis, results may be misinterpreted. The difficulty in identifying substances with LC-MS^n techniques is that their fragmentation pattern can only indicate whether a compound is conjugated on either the A- or the B-ring. Ion mobility spectrometry (IMS) may be a significant step forward in this regard. Ion mobility measurements allow ions to be separated from each other in a drift tube by a drift gas in an electric field. The ions need specific times to pass the tube, which are referred to as drift times and which depend on the instrument settings, the structure of the molecule and the mass-to-charge ratio. Independent collision cross section (CCS) values can be calculated for each individual ion. Thus, substances with the same *m/z*, such as isomers, can be distinguished on the basis of their CCS values.[7] However, as long as there are no sufficiently large databases for CCS values, standard substances are also required here for unambiguous identification. Although the use of online NMR techniques to study the metabolism seems to be possible,[8] a satisfactory and universally applicable method to accurately distinguish the position is feasible only with fully identified reference substances. In addition to the identification and quantification of phase II metabolites in biological samples, these compounds may be used to study their stability, distribution or kinetics in vitro. In most studies describing these parameters, often only the aglycone or sugar conjugates but not the metabolites of the polyphenols are used.[9] It is likely that depending on the conjugation position, cyanidin glucuronides will be metabolized differently, which is well-known for quercetin-7- and quercetin-4′-glucuronides in hepatocytes.[10] The metabolites may have a different biological activity and may not even enter the cells.[11] These facts should increasingly be considered and should lead to studies on the mode of action, for example, with respect to cell metabolism.[12] Various approaches have been described to synthesize phase II metabolites of flavonoids such as epicatechin or quercetin.[13,14] These routes include chemical ways, which may provide good yields but also give rise to a high number of isomers. Synthesis of metabolites via enzymes is advantageous in that complete regioselectivity is achieved and naturally occurring isomers are formed.

Enzymatic and chemical routes of synthesis of anthocyanin phase II metabolites have been described to a lesser extent. The stability of isolated anthocyanins and their metabolites is dependent on several factors such as light, pH value, temperature, and the presence of oxygen.[15] Therefore, a simple enzymatic synthesis of the metabolites is desirable. Concerning the in vitro metabolism of anthocyanins, existing literature refers particularly to the metabolism through human or rat enzymes. Although pigs are a widely accepted model in nutrition studies because they undergo a similar nutrient absorption process as humans,[16] pig liver has largely been unexplored in this field of research.

It is known from studies of Wu et al.[17] that cyanidin and its glucoside are metabolized to monoglucuronides and methylated conjugates in weanling pigs, whereas no information on sulfated derivatives was provided. In this work, the described in vivo metabolism[5] was simulated by the use of porcine liver as the enzyme source. The direction of a respective phase II reaction is determined through the addition of an activated cofactor in excess. These cofactors are 3′-phosphoadenosine-5′-phosphosulfate for sulfation, S-adenosyl-L-methionine for methylation, and uridine 5′-diphosphoglucuronic acid for glucuronidation. They were incubated with the substrate in the presence of the enzymes of a protein fraction. In this way, corresponding phase II metabolites should become accessible for characterization without the need of a previous concentration. Synthesized metabolites were characterized via LC-MS^n and LC-IMS-QTOF-MS.

2 Materials and methods

2.1 Chemicals

LC-MS-grade water, LC-MS-grade methanol, acetonitrile, and formic acid were purchased from ChemSolute (Renningen, Germany). Tris(hydroxymethyl)-aminomethane (Tris) (p.a.) was obtained from Merck (Darmstadt, Germany). Ethyl acetate was from VWR (Mannheim, Germany). Quercetin, uridine 5'-diphosphoglucuronic acid (UDPGA) trisodium salt (98-100%),

3'-phosphoadenosine-5'-phosphosulfate (PAPS) lithium salt (75%), S-adenosyl-L-methionine (SAM) chloride-dihydrochloride salt (75%), (−)-epicatechin (97%), and protocatechuic acid were purchased from Sigma-Aldrich (St. Louis, MO). Quercetin-3-*O*-glucopyranoside and (−)-epicatechin-3-*O*-gallate were from Extrasynthese (Genay, France). Dithiothreitol (DTT) (p.a.) was from Carl Roth (Karlsruhe, Germany). Blackberry concentrate was supplied by Haus Rabenhorst O. Lauffs GmbH & Co. KG (Unkel, Germany).

2.2 Isolation of anthocyanins

Diluted blackberry concentrate was purified from sugars and other polar compounds with a XAD7 HP resin (Sigma Aldrich, Munich, Germany). To separate anthocyanins from polyphenols, the procedure described by Juadjur and Winterhalter[18] was applied. The main anthocyanin in this extract was cyanidin-3-*O*-glucoside (**Figure 1**). To obtain the cyanidin aglycone, the lyophilized extract was treated with concentrated HCl for 90 min at 90 °C and afterwards extracted with ethyl acetate.

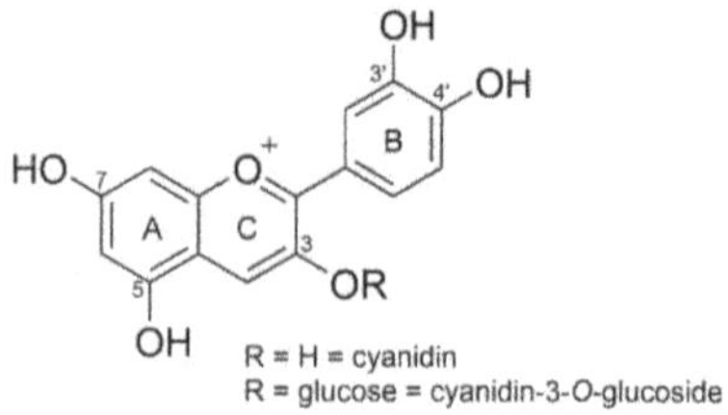

R = H = cyanidin
R = glucose = cyanidin-3-O-glucoside

Figure 1: Structure of cyanidin-3-*O*-glucoside and cyanidin

2.3 Preparation of liver fractions

Liver samples from pigs of the breed *Deutsche Landrasse* were obtained from the Institute of Animal Science (University of Bonn, Germany). After slaughter, all samples were frozen with liquid nitrogen and kept at −80 °C before analysis. Preparation of the different liver fractions was performed following a procedure reported by Rasmussen et al.[19]. Briefly, porcine liver was homogenized in about threefold amount of Tris-sucrose-buffer (10 mM Tris-HCl, 250 mM sucrose, pH 7.4) under ice cooling. To obtain the S9 fraction, the above-mentioned homogenate was centrifuged at 9,000 *g* for 10 min at 4 °C. The supernatant, the so-called S9 fraction, was collected. Subsequently, the S9 fraction was diluted by addition of precipitation buffer (10 mM Tris-HCl, 250 mM sucrose, 8 mM $CaCl_2$, pH 7.4) and again centrifuged (30 min, 18 000 *g*, 4 °C). The supernatant, the cytosolic fraction, was collected. The resulting microsomal pellets were resuspended in Tris-buffer (250 mM Tris-HCl, 1 mM EDTA, 3 M glycerol, pH 7.4). Each fraction was stored at −80 °C prior to analysis. Protein contents were determined as described by Bradford[20], with bovine serum albumin as standard.

2.4 Incubation

The reactions are based on the procedures developed by Fernandes et al.[21] and Vaidyanathan and Walle[14] and were optimized individually for each of the reactions. The final conditions are listed in **Table 1**. In a total volume of 250 µL of buffer, the substrate was incubated with the S9, microsomal or cytosolic fractions, respectively. The incubation mixture contained 4 mg protein/mL, $MgCl_2$, DTT, and the activated cofactor associated with the corresponding reaction (UDPGA, SAM or PAPS). Finally, depending on the solubility of the substrate, 400 µM of the substrate dissolved in buffer or DMSO was added. The mixture was incubated at 37 °C for 120–330 min under constant gentle shaking. The reaction was stopped by adding 50 µL of cold methanol to the mixture to precipitate proteins. After centrifugation (5 min, 7000 *g*) and microfiltration (regenerated cellulose, 0.20 µm), the supernatants were analyzed using LC-MS^n and LC-IMS-QTOF-MS. To ensure the activity of the enzymes and the reagents, especially of the activated cofactors, a positive control was carried out each day of measurement. Since it is known that the reaction works with rat liver enzymes,[22,14] the following substrates were used for the positive controls: protocatechuic acid for the methylation, quercetin and quercetin-3-*O*-glucoside for the glucuronidation, and (−)-epicatechin and epicatechin gallate for the sulfation reaction. Negative controls were carried out by omitting the activated cofactor.

Table 1: Parameters of the Incubation Reactions[a]

Reaction	Substrate	t (min)	Protein-fraction	$MgCl_2$ (mM)	DTT (mM)	Activated cofactor	Buffer
Sulfation							
	cy	120	cytosolic	5	8	0.1 mM PAPS	33 mM Tris-HCl
	cy3glu	120	S9	5	8	0.1 mM PAPS	
Methylation							
	cy	120	S9	1.2	1	2.5 mM SAM	33 mM phosphate
	cy3glu	190	S9	1.2	1	2.5 mM SAM	
Glucuronidation							
	cy	330	microsomal	100	-	2.5 mM UDPGA	10 mM phosphate
	cy3glu	330	microsomal	100	-	2.5 mM UDPGA	
Combination							
	cy	120	S9	1.2	1	0.1 mM PAPS, 2.5 mM SAM, 2.5 mM UDPGA	33 mM phosphate
	cy3glu	120	S9	1.2	1		

[a]pH value: 6; temperature: 37 °C, total volume: 250 µL; cy, cyanidin; cy3glu, cyanidin-3-*O*-glucoside

2.5 LC-MS-Analysis

UHPLC analysis of the reaction products was performed on an Acquity UPLC I-Class system (Waters, Milford, MA) consisting of a binary pump, an autosampler cooled at 10 °C, a column oven set at 40 °C, and a diode array detector scanning from 250 to 650 nm. An Acquity HSS-T3 RP18 column (150 mm x 2.1 mm; 1.8 μm particle size) combined with a pre-column (Acquity UPLC HSS T3 VanGuard, 100 Å, 2.1 mm x 5 mm, 1.8 μm), both from Waters (Milford, MA) was used for separation with water (A) and acetonitrile (B) as eluents, both acidified with 0.1% (v + v) formic acid. The flow rate was set at 0.4 mL/min. Analyses of the sulfation reaction of cyanidin were carried out with linear gradient conditions from 8% B to 10% B for 5 min, then to 17% B for 6 min, to 30% B for 4 min, and to 50% B for 8 min. The other reactions were analyzed using a gradient whose solvent composition changed between 0 and 20 minutes from 1 to 100% B with a concave gradient curve (Waters gradient profile 8). The injection volume was 5 μL. For MS analysis, the UHPLC was coupled with a LTQ-XL ion trap mass spectrometer (Thermo Scientific, Inc., Waltham, MA) equipped with an electrospray interface operating in positive ion mode for the anthocyanins and in negative ion mode for the positive controls. Ion mass spectra were recorded in the range of *m/z* 160–1200. The capillary was set at 325 °C with a spray voltage of 16 V for ESI^+, and at 350 °C and a spray voltage of −40 V for ESI^-. The source voltage was maintained at 4 (3) kV (ESI^-) at a current of 100 μA. The tube lens was adjusted to 55 V for ESI^+ and −55 V for ESI^-. Nitrogen was used as sheath, auxiliary and sweep gas at a flow of 70, 10 and 1 arb, respectively. Three consecutive scans were conducted: a full mass scan, a MS/MS scan of the most abundant ion of the first scan using normalized collision energy (CE) of 35%, and a MS^3 of the most abundant ion in the MS^2 with CE of 65%. To identify the generated conjugates, multiple reaction monitoring (MRM) measurements were performed and the masses resulting from typical losses of glucuronic acid $[M-176]^+/[M-H-176]^-$, a sulfate group $[M-80]^+/[M-H-80]^-$ and a methyl group $[M-15]^+/[M-H-15]^-$ were scanned. In recorded full scan measurements, the masses of the degradation products, such as protocatechuic acid and the corresponding aldehyde23, and their conjugates were additionally screened for.

For ion mobility spectrometry measurements, the UPLC was connected to a Vion IMS QTOF mass spectrometer (Waters, MA) operating in positive mode for anthocyanin containing samples and in negative mode for the positive controls. The capillary voltage was 0.5 kV for ESI^+ and 2.5 kV for ESI^-, the source temperature was 120 °C for ESI^+ and 100 °C for ESI^-, the cone voltage was 40 V, the desolvation gas temperature was 550 °C, and the desolvation gas flow was 1200 L/h (600 L/h for ESI^-). The measurements were conducted with automatic lock correction every 5 minutes with leucine-enkephaline as lock mass in a concentration of 100 pg/μL. Nitrogen was used as the drift gas and the MS mode was high definition with a low collision energy of 6 eV and a high collision energy ramp of 20–40 eV. Data were acquired and processed using UNIFI v1.9.2.045 (Waters, Milford, MA).

3 Results and discussion

Cyanidin-3-O-glucoside and cyanidin purified from blackberry juice and quercetin, quercetin-3-O-glucoside, (−)-epicatechin, (−)-epicatechin-3-O-gallate, and protocatechuic acid as positive controls were incubated with pig liver microsomal, cytosolic, or S9 fractions as a source of phase II enzymes. Activated cofactors were added to obtain methylated, glucuronidated or sulfated conjugates.

3.1 Methylation

Because catechol-O-methyltransferase (COMT) enzymes are located in the cytosol and bound to the membranes,[24] it was expected that the methylation reaction occurs within all liver fractions. Preliminary tests had shown that the most efficient fraction was the S9 fraction. On each day of measurement, a positive control was carried out. In the case of methylation, the substrate for the positive control was protocatechuic acid (PCA). It is a degradation product of cyanidin, and it has already been reported that PCA is methylated in vitro through rat liver enzymes.[22] To the best of our knowledge, this is the first work to show that PCA is also methylated by pig liver enzymes. Therefore, this compound was suitable as a positive control. The results of this reaction can be seen in **Tables 2** and **3**. After the methylation reaction of cyanidin-3-O-glucoside (**Figure 2**), two peaks with an absorption maximum at 520 nm were detected. Their mass spectra showed that one of them was the substrate cyanidin-3-O-glucoside with a molecular ion of *m/z* 449 ($[M]^+$), the other had a molecular mass of *m/z* 463 ($[M]^+$). The latter dissociated to *m/z* 301 ($[M-162]^+$) and 286 ($[M-162-15]^+$), which corresponds to a methylated derivative of cyanidin-3-O-glucoside. For cyanidin-3-O-glucoside, IMS measurements showed an accurate mass of *m/z* 449.1078 ($[M]^+$) with a CCS value of 201.3 ± 0.8 Å^2 and for the methylated derivative a *m/z* of 463.1236 ($[M]^+$) with a CCS value of 206.4 ± 0.3 Å^2. Since methylation enlarges the molecule, it is obvious that the CCS value also increases. The database MetCCS allows a prediction of CCS values of substances listed in the Human Metabolome Database (HMDB).[25] The CCS values calculated in this way are 201.6 Å^2 for cyanidin-3-O-glucoside and 203.2 Å^2 for peonidin-3-O-glucoside. Causon et al.[26] determined the CCS values of metabolites in red wine and found 207.1 Å^2 for peonidin-3-O-glucoside. These values are all consistent and show that CCS values are a further step toward the identification of a substance. Wu et al.[17] assumed that methylation is the preferred metabolic pathway of cyanidin-3-O-glucoside in pigs. It is conceivable that through methylation, the permeability of anthocyanins through membranes is increased.[27] In this way, they can be distributed more efficiently. For COMT enzymes, a catecholic structure is required for methylation.[24] Cyanidin-3-O-glucoside contains this structural element through the hydroxyl groups in the B ring. For this reason, methylation was most likely at these positions. As a result of the monomethylation of cyanidin-3-O-glucoside at position 3′ or 4′, the catecholic structure, which is a prerequisite for the COMT enzymes, got lost. By comparison of the retention time with commercially obtained peonidin-3-O-glucoside, it was ascertained that the pig liver enzymes introduced a methyl group at position 3′ of cyanidin-3-O-glucoside. Therefore, it is reasonable to assume that cyanidin-3-O-glucoside was not conjugated to more than one methyl group during incubation with the S9 fraction.

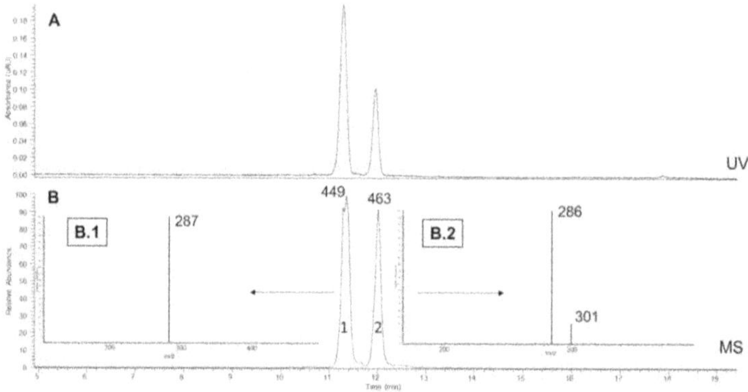

Figure 2: UHPLC-MS chromatograms of cyanidin-3-O-glucoside after methylation. A, UV chromatogram recorded at 521 nm. B, EIC of the *m/z* 449 ([M]$^+$, cyanidin-3-O-glucoside; peak 1) and the *m/z* 463 ([M]$^+$, methylated cyanidin-3-O-glucoside; peak 2) with the respective mass spectra (B.1 and B.2).

The results of the methylation experiments in this work differ from those reported by Wu et al.[17,28,29] In their in vivo studies, these authors described two different methylated cyanidin-3-O-glucosides in the urine of pigs. Both have the same molecular weight of *m/z* 463 ([M]$^+$) and the same product ion of *m/z* 301. One was confirmed as peonidin-3-O-glucoside by the retention time and comparison with a standard, the other one was named isopeonidin glucoside. However, the proposed structure was not confirmed by NMR spectroscopy. In addition, the presence and concentration of the metabolites were dependent on the composition of the anthocyanins in the fed berries. For example, the lack of methylated products from cyanidin-3-O-glucoside in the urine from chokeberry fed pigs was attributed to the increased competition for reaction sites on COMT by cyanidin-3-O-galactoside and cyanidin-3-O-arabinoside, which were present at much higher concentrations.[17] Fernandes et al.[21] achieved similar results in in vitro experiments. With the rat liver cytosolic protein fraction, two methylated metabolites of cyanidin-3-O-glucoside were synthesized. As the authors reported very close retention times, the gradient used in this work might have caused coelution of isopeonidin-3-O-glucoside and peonidin-3-O-glucoside. It should also be kept in mind that the enzymatic activities of pigs may change through their life and may be different among the breeds. Methylation reaction was also performed with cyanidin, but no methylated derivative was detected. One reason might be the instability of the cyanidin aglycone under physiological conditions.[30] As cyanidin-3-*O*-glucoside reaches the intestine, it gets cleaved. The released aglycone is then glucuronidated or again converted to cyanidin-3-*O*-glucoside.[5] Another reason might be the polarity of cyanidin. Passamonti et al.[27] reported that the aglycone was not a good substrate for the membrane carrier bilitranslocase, which is located in the liver plasma membrane and in epithelial cells of the gastric mucosa. Therefore, it is possible that cyanidin was not a substrate for the COMT located in the liver. In the context of the negative controls, it was observed that the content of cyanidin-3-*O*-glucoside was reduced during the incubation experiments to an average of two-third of the initial quantity, which may be caused by decomposition of the anthocyanin through enzymes, high pH values, and relatively high temperatures over long periods of time. Methylation was the only reaction that allowed monitoring of the results by UV spectroscopy. The maximum achievable amount of peonidin-3-*O*-glucoside was about onequarter of the initial amount of cyanidin-3-*O*-glucoside.

Table 2: LC-MS Identification of Enzymatically Synthesized Metabolites Using Cyanidin, Cyanidin-3-*O*-glucoside, (−)-Epicatechin, (−)-Epicatechin-3-*O*-gallate, Quercetin, Quercetin-3-*O*-glucoside and Protocatechuic Acid as Substrates

Compound	t_R *(MS)* *(min)*	*[M]*[+] *(m/z)*	*[M−H]*[−] *(m/z)*	*Fragment ions* MS^2*(m/z)*	*Fragment ions* MS^3 *(m/z)*[a]
Cyanidin					
	12.8	287		**213** (100), 231 (74.4), 241 (63.9), 259 (59.2)	154 (37.3), 157 (66.7), 185 (100)
Sulfated	12.3	367		**287** (100), 289 (38.3), 326 (11.3), 349 (23.0)	
	13.1	367		**287** (100), 289 (12.0)	
	15.9	367		**287** (100)	
Glucuronidated	11.2	463		287	
	12.0	463		287	
	12.2	463		287	
	14.0	463		287	
Methylated and glucuronidated	11.9	477		301	286
	12.2	477		301	286
	14.3	477		301	286
Sulfated and glucuronidated	14.9	543		**367** (100), 437 (47.7), 463 (30.5), 534 (54.5)	
Cyanidin-3-O-glucoside					
	11.3	449		287	213 (100), 231 (69.1), 241 (47.2), 259 (77.5)
Methylated	12.0	463		301	286
Sulfated	10.2	529		287 (93.9), **449** (100), 511 (4.0), 520 (6.8)	287
	11.1	529		**287** (100), 496 (10.3), 520 (19.5), 523 (13.7)	179 (76.5), 217 (100), 241 (61.7), 259 (73.0)
	11.7	529		**287** (100), 449 (31.5), 520 (12.8), 523 (9.8)	137 (50.0), 203 (100), 231 (77.2), 259 (81.7)
Glucuronidated	8.1	625		287 (97.5), 445 (4.2) 449 (26.2), **463** (100)	287 (100), 445 (1.2)
	9.6	625		287 (33.1), 449 (5.8), **463** (100)	287 (100)
	10.5	625		287 (7.6), 449 (1.8), **463** (100)	287 (100)
Methylated and glucuronidated	9.2	639		301 (48.3), 463 (40.1), **477** (100)	
	9.5	639		301 (43.8), 463 (34.6), **477** (100)	

	10.0	639		301 (43.9), 463 (16.6), **477** (100)	301
	10.8	639		301 (11.0), 463 (6.8), **477** (100)	301
(−)-Epicatechin					
	15.6		289	179 (11.5), 205 (31.8), **245** (100)	161 (19.8), 187 (26.2), 203 (100), 227 (22.6)
Methylated	19.3		303	217 (45.4), 244 (53.5), **259** (100), 285 (61.1)	217 (12.1), 244 (100)
	20.7		303	**137** (100), 244 (37.5), 259 (81.4), 285 (53.4)	83 (42.0), 93 (100), 109 (77.6)
Sulfated	13.3		369	137 (10.7), 217 (39.2), **289** (100)	179 (12.0), 205 (37.8), 245 (100)
	15.4		369	231 (72.1), **289** (100)	179 (10.6), 205 (45.3), 231 (10.8), 245 (100)
Glucuronidated	12.7		465	175 (15.4), **289** (100)	179 (13.2), 205 (44.1), 245 (100)
Methylated and glucuronidated	14.4		479	175 (64.8), **303** (100), 341 (23.0), 461 (43.9)	219 (67.4), 244 (51.4), 259 (100), 285 (75.6)
	15.4		479	289 (11.5), **303** (100), 313 (39.9), 447 (13.6)	217 (64.4), 235 (100), 244 (54.3), 259 (95.2)
	16.2		479	303	217 (55.7), 244 (45.8), 259 (100), 285(58.3)
	17.3		479	175 (95.4), **303** (100), 383 (24.7), 461 (61.9)	
	17.5		479	175 (57.5), 303 (66.7), 417 (11.7), **461** (100)	
	17.7		479	**175** (100), 303 (89.2), 411 (11.7), 461 (24.4)	
Methylated and sulfated	16.1		383	245 (75.7), **303** (100)	217 (49.2), 219 (65.5), 244 (57.6), 259 (100)
	17.2		383	137 (19.1), 217 (51.4), **303** (100), 304 (12.0)	219 (63.6), 244 (63.6), 259 (100), 285 (66.8)
	18.0		383	137 (11.4), 217 (24.6), **303** (100), 304 (17.7)	217 (44.4), 244 (59.9), 259 (100), 285 (41.8)
	18.8		383	137 (27.2), 217 (62.0), **303** (100), 304 (13.0)	137 (100), 244 (73.1), 259 (59.6), 285 (59.9)
	19.5		383	137 (27.9), 217 (47.2), **303** (100)	137 (89.8), 217 (31.5), 259 (100), 285 (70.9)
Quercetin					
	15.4		301	151 (87.7), **179** (100), 257 (18.2), 273 (20.0)	151
Methylated	16.3		315	300	151 (84.3), 255 (38.5), 271 (75.7), 272 (100)
Sulfated	15.0		381	301	151 (80.7), 179 (100), 257 (15.8), 273 (14.3)
Glucuronidated	13.4		477	301	151 (74.2), 179 (100), 257 (12.0), 273 (12.4)
	14.0		477	301	151 (60.0), 179 (100), 257 (14.9), 273 (12.9)
	14.3		477	301	151 (60.0), 179 (100), 257 (14.9), 273 (12.9)
Twice glucuronidated	13.0		653	477	301

Methylated and sulfated	15.0	395	315	151 (27.4), 300 (100)
	15.1	395	315	300
Methylated and glucuronidated	14.0	491	315	300
	14.1	491	315	300
	14.2	491	315	300
	14.4	491	315	300
	14.6	491	315	300
Methylated, glucuronidated and sulfated	16.5	571	315 (22.3), 395 (5.0), **491** (100)	
Quercetin-3-O-glucoside				
	13.3	463	301	151 (80.7), 179 (100), 193 (17.0), 273 (20.9)
Methylated	14.1	477	299 (73.7), **314** (100), 315 (61.3), 449 (64.9)	299
Twice methylated	14.5	491	300 (5.1), **315** (100), 329 (2.6), 473 (4.9)	300
Sulfated	13.3	543	301 (8.8), 381 (15.0), **463** (100)	301
Glucuronidated	12.4	639	301 (68.3), 459 (23.0), **463** (100),477 (50.1)	301
Methylated and glucosidated	13.5	639	300 (32.5), **315** (100)	300
	13.8	639	300 (33.5), **315** (100)	300
	13.9	639	300 (40.8), **315** (100)	300
Sulfated and methylated	13.0	557	241 (4.1), 315 (64.0), 395 (21.4), **477** (100)	299, 315, 357, 449
	13.2	557	241 (3.9), 315 (47.7), 395 (23.2), **477** (100)	299 (57.7), 314 (95.1), 315 (100), 449 (61.4)
	13.5	557	315 (22.3), **395** (100), 477 (20.7)	315
Methylated and glucuronidated	12.0	653	477	299 (75.3), 314 (100), 357 (40.2), 449 (57.9)
	12.7	653	315 (46.8), 473 (47.6), 477 (68.9), **491** (100)	315
	12.8	653	n.d.	
	12.9	653	477 (9.4), **491** (100), 533 (28.7)	315
Protocatechuic acid				
	7.3	153	109	
Methylated	11.5	167	108 (10.1), 123 (100), 152 (84.9), 167 (2.4)	
	11.9	167	108 (1.0), 117 (2.0), 123 (47.0), 152 (100)	

Sulfated	7.9	233	97 (6.8), 109 (3.2), **153** (100), 189 (79.3)	109 (100), 153 (10.5)
	8.3	233	97 (6.0), 109 (2.6), **153** (100), 189 (68.7)	109 (100), 153 (8.9)
Glucuronidated	6.1	329	113 (43.2), **153** (100) 175 (92.9), 193 (56.6)	109 (100), 153 (11.8)
	8.3	329	113 (35.5), **153** (100), 175 (51.1), 285 (27.8)	109
Sulfated and methylated	10.2	247	80 (2.6), 97 (7.7), **167** (100), 203 (60.2)	108 (15.3), 123 (85.8), 152 (100), 167 (11.0)
Methylated and glucuronidated	8.7	343	113 (63.1), 167 (18.9), **175** (100), 325 (2.0)	85 (5.7), 103 (5.0), 113 (100) 157 (6.4)
	10.2	343	113 (51.0), 123 (3.2), **167** (99.0), 175 (100)	108 (2.6), 123 (100) 152 (18.2), 167 (16.8)
	10.6		113 (93.8), 167 (13.7), 175 (100), 297 (65.9)	
	10.9		113 (66.6), 167 (51.9), **175** (100), 297 (15.1)	95 (3.9), 103 (7.1), 113 (100), 157 (4,6)
Epicatechin-3-O-gallate				
	13.7	441	169 (24.8), 271 (9.8), **289** (100), 331 (15.8)	179 (11.1), 205 (33.1), 231 (6.8), 245 (100)
Sulfated	14.2	521	169 (7.9), 289 (42.7), 441 (100), 442 (9.3)	169 (18.9), 193 (13.7), 289 (100), 331 (24.3)
	14.7	521	169 (7.1), 271 (2.5), 289 (32.7), 441 (100)	169 (18.8), 271 (10.9), 289 (100), 331 (28.2)
Glucuronidated	12.6	617	289 (24.7), 391 (3.1), **441** (100), 599 (16.1)	169 (12.3), 271 (14.0), 289 (100), 331 (24.0)
	12.7	617	289 (20.3), **441** (100), 465 (10.2), 599 (17.6)	289 (100), 329 (62.6), 331 (41.9), 423 (30.4)
	13.1	617	289 (20.3), **441** (100), 465 (10.2), 599 (17.6)	169 (12.1), 271 (9.2), 289 (100), 331 (18.3)
Sulfated and methylated	14.4	535	271 (48.8), 393 (14.0), 397 (52.9), **455** (100)	271 (47.0), 305 (100), 345 (24.8), 411 (24.9)
Glucuronidated and methylated	13.2	631	271 (20.1), 305 (4.5), 393 (4.3), **455** (100)	271 (99.1), 305 (100), 393 (23.8), 411 (48.3)
	13.4	631	271 (17.5), 359 (9.6), 411 (5.5), **455** (100)	271 (91.6), 305 (100), 345 (25.2), 411 (34.2)
	13.6	631	271 (18.5), 411 (7.1), **455** (100), 457 (6.3)	271 (100), 305 (92.1), 345 (22.2), 411 (43.7)
	13.8	631	271 (4.7), 289 (36.1), **455** (100), 457 (16.0)	183 (9.1), 289 (100), 303 (7.7), 437 (5.3)
Glucuronidated and twice methylated	14.1	645	285 (23.2), 305 (32.1), **469** (100), 627 (24.5)	271 (36.5), 305 (86.8), 319 (100), 425 (62.0)
Sulfated and twice methylated	14.8	549	285 (40.8), 305 (35.1), 411 (100), 469 (26.6)	331

[a]Precursors in bold; Base peak intensity in brackets; n.d. not detected

3.2 Glucuronidation

Because uridine 5′-diphosphoglucuronosyltransferase (UGT) enzymes are membrane bound,[31] glucuronidation reaction should take place only with the liver fractions that contain microsomal enzymes. Quercetin has already been shown to be glucuronidated in vitro through pig liver enzymes.[32] Therefore, in the case of glucuronidation, the substrates used for the positive controls were quercetin and quercetin-3-*O*-glucoside. In the present study, quercetin-3-*O*-glucoside was monoglucuronidated at not less than three different sites, which results in *m/z* 639 ($[M-H]^-$). The quercetin aglycone was also monoglucuronidated at three different sites (*m/z* 477, $[M-H]^-$). Additionally, it was glucuronidated twice (*m/z* 653, $[M-H]^-$). As expected, all measurements demonstrated that glucuronidation rendered the substances more polar, since the elution on the reversed phase shifted to earlier retention times. Glucuronidation experiments showed that cyanidin-3-*O*-glucoside as well as cyanidin were glucuronidated by microsomal liver enzymes. During the reaction, three compounds with *m/z* 625 ($[M]^+$), corresponding to a glucuronidated cyanidin-3-*O*-glucoside, were detected (**Figure 3**). The molecular mass $[M]^+$ at *m/z* 625 yielded the product ions *m/z* 463 ($[M-162]^+$), which exhibited the loss of a hexose, *m/z* 449 ($[M-176]^+$), which corresponds to the loss of glucuronic acid, and *m/z* 287 ($[M-162-176]^+$), indicating the loss of both of the abovementioned moieties. Thus, monoglucuronidated cyanidin-3-*O*-glucosides were tentatively identified. The main fragments of the molecular ion $[M]^+$ 625 and the accurate masses are listed in **Tables 2** and **3**. It can also be seen that the conjugated forms eluted earlier from the C18 phase than the cyanidin-3-*O*-glucoside because they are more polar. The three glucuronides have CCS values of 252.8 ± 0.8 Å^2, 249.9 ± 0.3 Å^2, and 239.5 ± 0.5 Å^2 (listed in their elution order). As described for quercetin,[33] glucuronidation has a greater influence on the CCS value than methylation, the glucuronide with the smallest CCS value being that glucuronidated at the 3′-OH position. Since cyanidin glucuronides are not listed in the HMDB and thus some molecular descriptors are missing, a prediction of the CCS values with the MetCCS database is not possible. Due to the lack of published data, no statements on the conjugation positions (3′, 4′, 5, 7) can be made here. Likewise, the fragmentation pattern did not allow an assignment of the peaks to the possible isomers. In order to obtain typical fragments that indicate the position of the conjugation, the C ring of the cyanidin would need to be cleaved,[34] but this cleavage would necessitate a high collision energy. However, this leads to the cleavage of the weakest bonds, which are those between cyanidin and the glucoside and the glucuronide moiety. An increase in collision energy resulted in untypical fragments, which did not allow the unambiguous identification. Nevertheless, the three peaks described had qualitatively the same fragmentation pattern, but these differed in their intensities (**Figure 3**). Clifford et al.[35] showed that it is possible to distinguish between different isomers of chlorogenic acid by the intensity of the fragment masses. If the isomers of glucuronidated cyanidin-3-*O*-glucoside have been identified by, for example, chemically synthesized compounds, then it may be possible to differentiate between substances found in vivo based on their fragmentation patterns, their intensities and their CCS values.

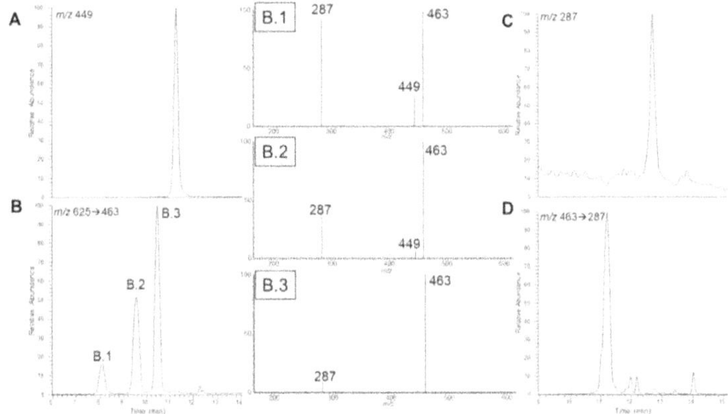

Figure 3: UHPLC-MS chromatograms of cyanidin-3-*O*-glucoside (A, B) and cyanidin (C, D) after glucuronidation. A: EIC of the *m/z* 449 ($[M]^+$, cyanidin-3-*O*-glucoside) B: EIC of the *m/z* 463 ($[M]^+$, glucuronidated cyanidin) of a precursor ion scan of the m/z 625 ($[M]^+$, glucuronidated cyanidin-3-*O*-glucoside) C: EIC of the *m/z* 287 ($[M]^+$, cyanidin) **D**: SRM chromatogram of the *m/z* 463 ($[M]^+$, glucuronidated cyanidin) to the *m/z* 287 ($[M]^+$, cyanidin)

In addition, cyanidin was glucuronidated through microsomal liver enzymes. The results are shown in **Figure 3**. Cyanidin is a relatively unstable compound that decomposes at elevated temperatures and high pH values.[36] Therefore, after incubation of the negative controls, cyanidin could no longer be found. In contrast, after incubation with the cofactors and thus after glucuronidation, cyanidin glucuronide was detected. It is well-known that glycosides of cyanidin are more stable[36] and it is likely that glucuronidation may have the same stabilizing effect. **Figure 3** shows the extracted ion chromatogram (EIC) of *m/z* 463 → 287 ($[M]^+$). At least four peaks were detected. Three of them eluted earlier than cyanidin, which is in accordance with their increasing polarity. The first peak has almost the same retention time as cyanidin-3-*O*-glucoside. Davis and Brodbelt[37] set up an elution order of different quercetin glucosides and glucuronides, wherein the 3-*O*-glucoside and the 3-*O*-glucuronide have approximately the same position. Because of the structural similarity of quercetin and cyanidin, and due to the similarity of the CCS values (201.3 ± 0.4 Å^2 for cyanidin-3-*O*-glucoside and 202.6 ± 0.4 Å^2 for cyanidin glucuronide), it may be assumed that the first peak corresponds to a cyanidin-3-*O*-glucuronide. However, one derivative eluted later than cyanidin; this may be caused by steric reasons. With this reaction, maximum yields of glucuronidated cyanidin-3-*O*-glucoside of about one percent of the applied cyanidin-3-*O*-glucoside were achieved. The yield of glucuronidated cyanidin could not be determined due to the small amounts and the instability of cyanidin.

Table 3: LC-IMS-QTOF-MS Analysis Data of Cyanidin, Cyanidin-3-*O*-glucoside, (−)-Epicatechin, (−)-Epicatechin-3-*O*-gallate, Quercetin, Quercetin-3-*O*-glucoside and Protocatechuic Acid Obtained after Incubation with Porcine Liver Enzymes

Compound	t_R *(min)*	*[M]⁺ (m/z)*	*[M−H]⁻ (m/z)*	*CCS value/[M]⁺ (Å²)*	*CCS value/[M−H]⁻ (Å²)*	*Mass error (mDa)*	*Mass error (ppm)*
Cyanidin							
	13.1	287.0550		162.8 ± 0.8		0.0	+0.1
Sulfated	11.3	367.0138		152.4 ± n.s.		+2.0	+5.5
Glucuronidated	11.6	463.0874		202.6 ± 0.4		+0.3	+0.7
	12.1	463.0902		172.6 ± n.s.		+3.0	+6.6
	12.6	463.0848		161.3 ± n.s.		−2.3	−5.0
Methylated and glucuronidated	12.1	477.1045		171.0 ± 1.1		+1.7	+3.6
	13.9	477.1027		164.5 ± 0.3		−0.2	−0.3
Cyanidin-3-O-glucoside							
	11.7	449.1078		201.3 ± 0.4		0.0	0.0
Methylated	12.3	463.1236		206.4 ± 0.3		0.0	+0.1
Sulfated	10.7	529.0654		213.1 ± 0.1		+0.8	+1.5
Glucuronidated	9.1	625.1402		252.8 ± 0.8		+0.2	+0.4
	10.2	625.1401		249.9 ± 0.3		+0.2	+0.3
	11.0	625.1398		239.5 ± 0.5		−0.2	−0.3
Methylated and glucuronidated	11.2	639.1583		245.9 ± 0.4		+2.6	+4.1
(−)-Epicatechin							
	12.5		289.0718		157.5 ± 0.1	0.0	0.0
Methylated	13.5		303.0870		168.6 ± n.s.	−0.4	−1.4
	13.9		303.0871		175.6 ± n.s.	−0.4	−1.2
Sulfated	11.9		369.0289		172.4 ± 0.3	−0.1	−0.4
	12.5		369.0283		176.3 ± 0.5	−0.2	−0.6
Glucuronidated	11.5		465.1037		196.9 ± n.s.	−0.2	−0.4
	12.1		465.1034		195.9 ± n.s.	−0.4	−0.9

Methylated and glucuronidated	11.9	479.1187	198.8 ± n.s.	−0.8	−1.7
	12.3	479.1194	205.2 ± n.s.	−0.1	−0.3
	12.5	479.1190	208.2 ± n.s.	−0.5	−1.1
	12.7	479.1194	203.9 ± n.s.	−0.1	−0.2
Methylated and sulfated	12.5	383.0437	191.6 ± n.s.	−0.5	−1.4
	12.9	383.0438	180.1 ± n.s.	−0.5	−1.4
	13.3	383.0437	179.9 ± n.s.	−0.4	−1.1
Quercetin					
	15.6	301.0350	161.6 ± 0.2	+0.1	+0.2
Methylated	16.2	315.0509	169.2 ± 0.4	0.0	0.0
Sulfated	15.4	380.9920	181.5± n.s.	+0.0	+0.1
Glucuronidated	13.6	477.0672	197.8 ± 0.0	+0.1	+0.3
	14.3	477.0670	161.5 ± 0.4	0.0	−0.1
	14.4	477.0670	210.5 ± 1.4	−0.1	−0.1
Twice glucuronidated	13.4	653.1002	239.1 ± 0.9	−0.3	−0.5
Methylated and sulfated	15.5	395.0078	188.4 ± n.s.	−0.2	−0.4
Methylated and glucuronidated	14.6	491.0828	217.2 ± n.s.	−0.5	−1.0
	14.6	491.0828	204.9 ± 1.1	−0.5	−1.0
Methylated, glucuronidated and sulfated	14.4	571.0392	214.9 ± n.s.	−0.8	−1.4
Quercetin-3-O-glucoside					
	13.7	463.0872	201.6 ± n.s.	−1.0	−2.1
Methylated	14.4	477.1032	208.1 ± n.s.	−0.7	−1.4
	15.2	477.1030	208.7 ± n.s.	−0.9	−1.9
Sulfated	12.4	543.0439	213.1 ± n.s.	−1.1	−2.1
	12.9	543.0437	222.7 ± n.s.	−1.3	−2.4
	13.6	543.0436	209.4 ± n.s.	−1.4	−2.6
Glucuronidated	11.6	639.1200	236.7 ± n.s.	−0.3	−0.5

	12.8	639.1191	231.8 ± n.s.	−1.2	−1.9
Methylated and glucuronidated	12.4	653.1349	241.8 ± n.s.	−1.1	−1.7
	13.0	653.1346	240.0 ± n.s.	−1.4	−2.2
	13.2	653.1348	243.1 ± n.s.	−1.2	−1.8
Sulfated and methylated	13.3	557.0594	220.1 ± n.s.	−1.3	−2.3
Protocatechuic acid					
	8.8	153.0192	121.5 ± 0.3	−0.2	−1.1
Methylated	12.0	167.0349	128.2 ± 0.2	−0.2	−1.0
Sulfated	7.8	232.9762	137.9 ± 0.2	0.0	+0.1
	8.2	232.9761	137.7 ± 0.3	0.0	−0.1
	8.7	232.9762	138.0 ± 0.2	+0.1	+0.3
Glucuronidated	7.4	329.0516	167.5 ± 0.6	+0.2	+0.5
	8.9	329.0512	164.0 ± 0.1	−0.2	−0.6
Sulfated and methylated	10.0	246.9917	144.2 ± 0.2	−0.1	−0.4
	12.8	246.9915	144.5 ± 1.2	−0.3	−1.2
Methylated and glucuronidated	9.2	343.0669	177.4 ± 0.3	−0.3	−0.7
	10.3	343.0668	169.1 ± 0.5	−0.3	−0.9
	10.7	343.0675	166.6 ± 0.5	+0.4	+1.1
(−)-Epicatechin-3-O-gallate					
	13.7	441.0824	199.0 ± n.s.	−0.3	−0.7
Methylated	14.3	455.0980	204.1 ± n.s.	−0.4	−0.8
Sulfated	13.8	521.0391	205.1 ± n.s.	−0.5	−0.9
	14.2	521.0390	204.0 ± n.s.	−0.5	−0.9
	14.3	521.0390	205.1 ± n.s.	−0.5	−0.9
Glucuronidated	12.6	617.1151	230.8 ± n.s.	+0.3	+0.5
	13.1	617.1145	217.9 ± n.s.	−0.3	−0.5
	13.6	617.1150	230.1 ± n.s.	+0.2	+0.3

	13.8	617.1120	236.2 ± n.s.	−2.8	−4.6
Sulfated and methylated	14.1	535.0547	207.5 ± n.s.	−0.5	−0.9
Methylated and glucuronidated	13.2	631.1300	241.5 ± n.s.	−0.5	−0.8
	13.4	631.1298	227.6 ± n.s.	−0.7	−1.1
	13.6	631.1303	224.3 ± n.s.	−0.2	−0.3
Sulfated and twice methylated	14.5	549.2840	207.1 ± n.s.	−0.6	−1.1
Glucuronidated and twice methylated	13.4	645.1459	244.3 ± n.s.	−0.2	−0.3
	14.0	645.1464	246.6 ± n.s.	+0.3	+0.4
	14.2	645.1463	234.1 ± n.s.	+0.2	+0.4

n.s. not specified

3.3 Sulfation

Sulfotransferases (SULT) exist both as cytosolic and as membrane-bound enzymes.[38] Thus, the sulfation reaction may work with all obtained liver fractions. To monitor the reaction conditions, (−)-epicatechin and (−)-epicatechin-3-*O*-gallate were used as substrates for the positive control. The successful sulfation of these substrates with rat liver has already been described.[14,39] Also in this work, two differently substituted isomers of monosulfated (−)-epicatechin, one at the A-ring and one at the B-ring, were detected. For (−)-epicatechin gallate the conjugation position remains unclear due to inconclusive fragmentation. Under the tested in vitro conditions, sulfated metabolites of cyanidin-3-*O*-glucoside were obtained with the S9 fraction and of the cyanidin aglycone with the cytosolic fraction. Presumably, the SULT enzymes of the porcine liver are able to bind cyanidin-3-*O*-glucoside and cyanidin as a substrate to transfer the sulfonate group. The presence of the sulfated cyanidin-3-*O*-glucosides was confirmed by detection of the precursor at *m/z* 529 ($[M]^+$) and product ions at *m/z* 449 and *m/z* 287 (**Figure 4**). All three peaks of $[M]^+$ 529 decay to a fragment at *m/z* 287, which corresponds to $[M-162-80]^+$. A *m/z* value of 80 for the substitution group is indicative of a sulfate residue. However, it should be noted that the loss of 80 u was not observed in the peak that elutes second.

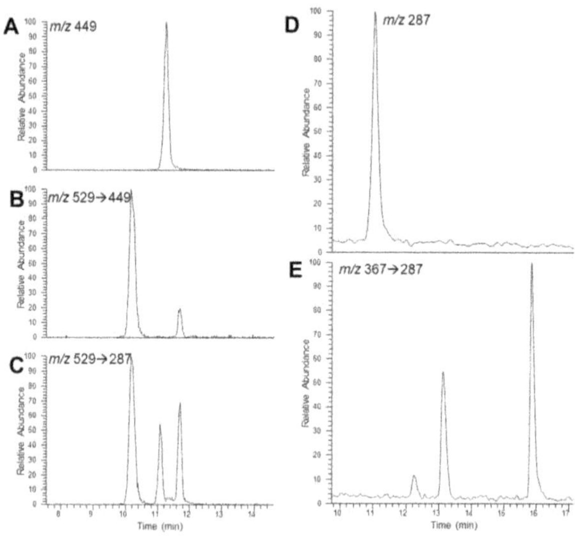

Figure 4: UHPLC-MS chromatograms of cyanidin-3-*O*-glucoside (A,B,C) and cyanidin (D,E) after sulfation. A: EIC of the *m/z* 449 ($[M]^+$, cyanidin-3-*O*-glucoside) B: EIC of the *m/z* 449 ($[M]^+$, cyanidin-3-*O*-glucoside) of a precursor ion scan of the *m/z* 529 ($[M]^+$, sulfated cyanidin-3-*O*-glucoside) **C**: EIC of the *m/z* 287 ($[M]^+$, cyanidin) of a precursor ion scan of the *m/z* 529 ($[M]^+$, sulfated cyanidin-3-*O*-glucoside) **D**: EIC of the *m/z* 287 ($[M]^+$, cyanidin) **E**: SRM chromatogram of the *m/z* 367 ($[M]^+$, sulfated cyanidin) to the *m/z* 287 ($[M]^+$, cyanidin)

This might imply that the second peak belongs to a cyanidin glucoside sulfated at the sugar moiety, but the exact site of sulfation could not be specified for any of these compounds. The IMS measurements of the sulfation reaction of cyanidin-3-*O*-glucoside resulted in only one assignable derivative. Its accurate mass was 529.0654 ($[M]^+$) and the CCS value was 213.1 ± 0.1 Å^2. As already shown by Chalet et al.[33] for epicatechin and epicatechin-3-*O*-gallate, the CCS value is increased by sulfation more than by methylation, but not as much as by glucuronidation. The values of the positive controls with epicatechin and epicatechin gallate (**Table 3**) are also remarkably consistent with those calculated by the MetCCS predictor and those described by Chalet et al.[33] Also, the incubation of the cytosolic liver fraction with cyanidin was successful, as sulfation at three different positions was observed via LC-MS^n. In the selected reaction monitoring (SRM) chromatograms, the transition of *m/z* 367 ($[M]^+$) to *m/z* 287 at three retention times (12.2, 13.1, and 15.8 min) was observed. This difference of 80 u describes the cleavage of a sulfate group from a cyanidin sulfate. Cyanidin (t_R = 11.0 min) was not detected after the incubation. The IMS measurements can also be referenced to only one derivative in case of the sulfation of cyanidin. The accurate mass of this sulfate is 367.0138 ($[M]^+$) and the CCS value is 152.4 Å^2. Since only one measurement was carried out, no uncertainty can be specified here. The smaller CCS value in comparison to the initial substance is conspicuous and can only be attributed to a different conformation of the molecule through conjugation. The position of the sulfate group could not unambiguously be determined by means of the fragmentation pattern or IMS data. These findings show that the method used is well suited to obtain sulfated derivatives of anthocyanins. The data obtained may be used to build up a database and thus to support studies on, e.g., the bioavailability of phenolic compounds. The results are contradictory to the results of the in vivo study of Wu et al.,[28] who did not detect any sulfated anthocyanins. A possible reason is the use of a C18 SPE cartridge for sample preparation by these authors. As sulfated substances are more polar than their nonsulfated derivatives, they might have passed the stationary phase without retention. This may imply that the extraction method chosen by Wu et al.[28] was sufficient for the other metabolites but did not retain the sulfated compounds. Furthermore, a study conducted by Feliciano et al.[40] showed that the recovery of sulfated metabolites on a HLB stationary phase was rather poor and thus any sulfated compounds that may have formed remained undetected by the selection of a C18 phase. In this work, approximately 0.06% of the applied cyanidin-3-*O*-glucoside was sulfated. The yield of the sulfation reaction of cyanidin could not be determined because of the small amounts and the instability of cyanidin.

3.4 Multiple Conversions

To investigate whether also mixed conjugated derivatives were formed by the liver enzymes through simultaneous incubation with PAPS, SAM, and UDPGA, the reaction was conducted with the S9 fraction as the enzyme source. Thus, it was also possible to check whether preferential metabolic pathways or synergistic effects exist. In the S9 fraction, both cytosolic and microsomally bound enzymes are present. (−)-Epicatechin, quercetin-3-*O*-glucoside, protocatechuic acid, epicatechin-3-*O*-gallate, and quercetin served as positive controls. The resulting conjugates were characterized by their mass and fragmentation patterns (MRM/SRM transitions), which are listed in **Table 2**, and their accurate mass and CCS values, which are shown in **Table 3**. In the experiments with cyanidin-3-*O*-glucoside and cyanidin, the masses of a glucuronidated and methylated derivative of each compound were detected. In addition, the cyanidin aglycone was simultaneously glucuronidated and sulfated. The experiments showed that there is a correlation between the preferred metabolic pathway and the polarity of the substrate (**Table 4**). By methylation and, depending on the polarity of the substrate, another conjugation, a similar polarity of the conjugates was produced. Whereas the apolar aglycones, quercetin, cyanidin, and epicatechin were both methylated and glucuronidated, the more polar -3-*O*-glucosides were mainly methylated.

Table 4: Preferred Metabolic Pathway of Phenolic Substrates[a]

Substrate		*Preferred metabolic pathway*
Cyanidin-3-*O*-glucoside	↑ Solubility in water ↓	Methylation
Quercetin-3-*O*-glucoside		Methylation
(−)-Epicatechin-3-*O*-gallate		Methylation + sulfation
(−)-Epicatechin		Methylation + sulfation
Cyanidin		Methylation + glucuronidation
Quercetin		Methylation + glucuronidation

[a]pH-value in the liver extract: 6

Since this is also the case with cyanidin-3-*O*-glucoside, it can be concluded that methylation is the preferred metabolic pathway in porcine liver. Since the relatively long periods of incubation time and the elevated temperatures may promote the degradation of cyanidin and possibly also of its derivatives, these products were additionally searched for in the fullscan measurements. These screenings revealed the presence of protocatechuic acid in few of the samples, but no conjugated form of any degradation product was detected. However, it is difficult to make an exact statement since the liver is a very complex matrix and the fullscan mode may not be sensitive enough to detect degradation products that may have been produced. Nevertheless, this work advances research in the field of synthesis of metabolites because so far only targeted reactions in the direction of a certain metabolic pathway of anthocyanins have been carried out. Multiple conversions in vitro have not yet been described in detail. These, however, represent more realistic reactions taking place in the liver. In addition, these in vitro conversions are a simple way to generate metabolites, which may occur in vivo and to use them as reference data. The values listed in **Tables 2** and **3**, especially the correlation of mass spectra and CCS values, are so far unique and very helpful in the identification of ions by comparison with data acquired under comparable conditions. The results of the in vitro experiments carried out here are in partial agreement with those reported by Wu et al.[17] The authors also identified a monomethylated derivative of cyanidin-3-*O*-glucoside as the main metabolic product in pigs. Cyanidin-3-*O*-glucoside fed via chokeberry, elderberry, and blackcurrant was metabolized and two different monoglucuronides, one monomethylated derivative, and one methylated glucuronide were detected. In addition, one monoglucuronide and one methylglucuronide of cyanidin were formed. Because the berries did not contain cyanidin, the authors assumed that cyanidin was formed by endogenous glucosidases in the pig and was then metabolized. Sulfated metabolites were not reported.

The results of in vitro experiments cannot be directly transferred to the situation in vivo because the metabolism, genetic differences, and food matrices cannot be considered completely. It should also be taken into account that different methods were applied. On the one hand, different pig breeds were used and on the other hand, in the in vivo study the authors found differences in metabolism depending on the different berries. Moreover, the metabolites were determined in the urine of the pigs, whereas the metabolites derived from this study were identified directly from the liver matrix. The results of Fernandes et al.[21] are in better agreement with those obtained in this work. Through rat liver, four different glucuronides and two different methylated derivatives of cyanidin-3-*O*-glucoside were synthesized. Combination or sulfation experiments have not been conducted. De Ferrars et al.[41] fed ^{13}C-labeled cyanidin-3-*O*-glucoside to humans and found, among others, two isomers of cyanidin glucuronide, peonidin glucoside, one methylated cyanidin glucuronide, and three methylated cyanidin-3-*O*-glucoside glucuronides, the latter occurring predominantly. Although this was an in vivo experiment in humans, it is surprisingly consistent with the results obtained in this work.

Some limitations of the described experiments need to be discussed. It should be considered that the metabolites formed were very unstable, and the samples had to be analyzed directly after the experiment. Another fact that should not be ignored is that the enzymes are not equally distributed in the liver cells, which may compromise the reproducibility. When multiple conjugates resulted from the reaction owing to the presence of multiple available hydroxyl groups, it was not possible to determine the exact position of the conjugation with mass spectrometry, and NMR measurements would need larger quantities of the metabolites. An exception was the methylation reaction of cyanidin-3-*O*-glucoside, where a peonidin-3-*O*-glucoside standard substance was available and the conjugation position could be determined.

Finally, it can be concluded that the approach for the synthesis of anthocyanin phase II metabolites presented in this work is suitable on a small scale. The reactions are limited in their volume through the amount of the liver and of the cofactors that are available. To obtain mass spectra and CCS values of the metabolites, to create a database, and to compare the retention times with biological samples, the described synthesis is a feasible way to synthesize the metabolites that are produced by human or porcine enzymes.

Funding

This research was supported by Diet Body Brain, Competence Cluster of Nutrition, funded by the German Federal Ministry of Education and Research (BMBF) (Grant No. 01EA1410A).

4 References

(1) Horbowicz, M.; Kosson, R.; Grzesiuk, A.; Dębski, H. Anthocyanins of fruits and vegetables – their occurrence, analysis and role in human nutrition. Veg. Crops Res. Bull. 2008, 68, p. 365.

(2) Clifford, M. N. Anthocyanins – nature, occurrence and dietary burden. J. Sci. Food Agric. 2000, 80, pp. 1063–1072.

(3) Pojer, E.; Mattivi, F.; Johnson, D.; Stockley, C. S. The case for anthocyanin consumption to promote human health. A review. Compr. Rev. Food Sci. Food Saf. 2013, 12, pp. 483–508.

(4) Cho, M. J.; Howard, L. R.; Prior, R. L.; Clark, J. R. Flavonoid glycosides and antioxidant capacity of various blackberry, blueberry and red grape genotypes determined by high-performance liquid chromatography/mass spectrometry. J. Sci. Food Agric. 2004, 84, pp. 1771–1782.

(5) Kay, C. D.; Pereira-Caro, G.; Ludwig, I. A.; Clifford, M. N.; Crozier, A. Anthocyanins and flavanones are more bioavailable than previously perceived. A review of recent evidence. Annu. Rev. Food Sci. Technol. 2017, 8, pp. 155–180.

(6) Passon, M.; Bühlmeier, J.; Zimmermann, B. F.; Stratmann, A.; Latz, S.; Stehle, P.; Galensa, R. Polyphenol phase-II metabolites are detectable in human plasma after ingestion of 13C labeled spinach-a pilot intervention trial in young healthy adults. Mol. Nutr. Food Res. 2018, 62, p. 1701003.

(7) Lapthorn, C.; Pullen, F.; Chowdhry, B. Z. Ion mobility spectrometry-mass spectrometry (IMS-MS) of small molecules: separating and assigning structures to ions. Mass Spectrom. Rev. 2013, 32, pp. 43–71.

(8) Lehtonen, H.-M.; Lindstedt, A.; Järvinen, R.; Sinkkonen, J.; Graça, G.; Viitanen, M.; Kallio, H.; Gil, A. M. 1H NMR-based metabolic fingerprinting of urine metabolites after consumption of lingonberries (Vaccinium vitis-idaea) with a high-fat meal. Food Chem. 2013, 138, pp. 982–990.

(9) Middleton, E.; Kandaswami, C.; Theoharides, T. C. The effects of plant flavonoids on mammalian cells: Implications for inflammation, heart disease, and cancer. Pharmacol Rev. 2000, 52, pp. 673–751.

(10) O'Leary, K. A.; Day, A. J.; Needs, P. W.; Mellon, F. A.; O'Brien, N. M.; Williamson, G. Metabolism of quercetin-7- and quercetin-3-glucuronides by an in vitro hepatic model: the role of human β-glucuronidase, sulfotransferase, catechol-O-methyltransferase and multi-resistant protein 2 (MRP2) in flavonoid metabolism. Biochem. Pharmacol. 2003, 65, pp. 479–491.

(11) Spencer, J. P.E.; Abd El Mohsen, M. M.; Rice-Evans, C. Cellular uptake and metabolism of flavonoids and their metabolites: implications for their bioactivity. Arch. Biochem. Biophys. 2004, 423, pp. 148–161.

(12) Aragonès, G.; Danesi, F.; Del Rio, D.; Mena, P. The importance of studying cell metabolism when testing the bioactivity of phenolic compounds. Trends Food Sci. Technol. 2017, 69, pp. 230–242.

(13) Needs, P. W.; Kroon, P. A. Convenient syntheses of metabolically important quercetin glucuronides and sulfates. Tetrahedron 2006, 62, pp. 6862–6868.

(14) Vaidyanathan, J. B.; Walle, T. Glucuronidation and sulfation of the tea flavonoid (-)-epicatechin by the human and rat enzymes. Drug Metab. Dispos. 2002, 30, pp. 897–903.

(15) Fossen, T.; Cabrita, L.; Andersen, O. M. Colour and stability of pure anthocyanins influenced by pH including the alkaline region. Food Chem. 1998, 63, pp. 435–440.

(16) Miller, E. R.; Ullrey, D. E. The pig as a model for human nutrition. Annu. Rev. Nutr. 1987, 7, pp. 361–382.

(17) Wu, X.; Pittman, H. E.; McKay, S.; Prior, R. L. Aglycones and sugar moieties alter anthocyanin absorption and metabolism after berry consumption in weanling pigs. J. Nutr. 2005, 135, pp. 2417–2424.

(18) Juadjur, A.; Winterhalter, P. Development of a novel adsorptive membrane chromatographic method for the fractionation of polyphenols from bilberry. J. Agric. Food Chem. 2012, 60, pp. 2427–2433.

(19) Rasmussen, M. K.; Ekstrand, B.; Zamaratskaia, G. Comparison of cytochrome P450 concentrations and metabolic activities in porcine hepatic microsomes prepared with two different methods. Toxicol. In Vitro 2011, 25, pp. 343–346.

(20) Bradford, M. M. A rapid and sensitive method for the quantitation of microgram quantities of protein utilizing the principle of protein-dye binding. Anal. Biochem. 1976, 72, pp. 248–254.

(21) Fernandes, I.; Azevedo, J.; Faria, A.; Calhau, C.; Freitas, V. de; Mateus, N. Enzymatic hemisynthesis of metabolites and conjugates of anthocyanins. J. Agric. Food Chem. 2009, 57, pp. 735–745.

(22) Fleschhut, J. Untersuchungen zum Metabolismus, zur Bioverfügbarkeit und zur antioxidativen Wirkung von Anthocyanen. Dissertation: University of Karlsruhe, Germany, 2004.

(23) Castañeda-Ovando, A.; Pacheco-Hernández, M. d. L.; Páez-Hernández, M. E.; Rodríguez, J. A.; Galán-Vidal, C. A. Chemical studies of anthocyanins: A review. Food Chem. 2009, 113, pp. 859–871.

(24) Cao, Y.; Chen, Z.-J.; Jiang, H.-D.; Chen, J.-Z. Computational studies of the regioselectivities of COMT-catalyzed meta-/para-O methylations of luteolin and quercetin. J. Phys. Chem. B 2014, 118, pp. 470–481.

(25) Zhou, Z.; Shen, X.; Tu, J.; Zhu, Z.-J. Large-Scale Prediction of Collision Cross-Section Values for Metabolites in Ion Mobility-Mass Spectrometry. Anal. Chem. 2016, 88, pp. 11084–11091.

(26) Causon, T. J.; Ivanova-Petropulos, V.; Petrusheva, D.; Bogeva, E.; Hann, S. Fingerprinting of traditionally produced red wines using liquid chromatography combined with drift tube ion mobility-mass spectrometry. Anal. Chim. Acta 2019, 1052, pp. 179–189.

(27) Passamonti, S.; Vrhovsek, U.; Mattivi, F. The interaction of anthocyanins with bilitranslocase. Biochem. Biophys. Res. Commun. 2002, 296, pp. 631–636.

(28) Wu, X.; Pittman, H. E.; Prior, R. L. Pelargonidin is absorbed and metabolized differently than cyanidin after marionberry consumption in pigs. J. Nutr. 2004, 134, pp. 2603–2610.

(29) Wu, X.; Pittman, H. E.; Prior, R. L. Fate of anthocyanins and antioxidant capacity in contents of the gastrointestinal tract of weanling pigs following black raspberry consumption. J. Agric. Food Chem. 2006, 54, pp. 583–589.

(30) Ichiyanagi, T.; Shida, Y.; Rahman, M. M.; Hatano, Y.; Matsumoto, H.; Hirayama, M.; Konishi, T. Metabolic pathway of cyanidin 3-O-beta-D-glucopyranoside in rats. J. Agric. Food Chem. 2005, 53, pp. 145–150.

(31) Jancova, P.; Anzenbacher, P.; Anzenbacherova, E. Phase II drug metabolizing enzymes. Biomed. Pap. 2010, 154, pp. 103–116.

(32) Dueñas, M.; Mingo-Chornet, H.; Pérez-Alonso, J. J.; Di Paola-Naranjo, R.; González-Paramás, A. M.; Santos-Buelga, C. Preparation of quercetin glucuronides and characterization by HPLC–DAD–ESI/MS. Eur. Food Res. Technol. 2008, 227, pp. 1069–1076.

(33) Chalet, C.; Hollebrands, B.; Janssen, H.-G.; Augustijns, P.; Duchateau, G. Identification of phase-II metabolites of flavonoids by liquid chromatography-ion-mobility spectrometry-mass spectrometry. Anal. Bioanal. Chem. 2018, 410, pp. 471–482.

(34) Cren-Olivé, C.; Déprez, S.; Lebrun, S.; Coddeville, B.; Rolando, C. Characterization of methylation site of monomethylflavan-3-ols by liquid chromatography/electrospray ionization tandem mass spectrometry. Rapid Commun. Mass Spectrom. 2000, 14, pp. 2312–2319.

(35) Clifford, M. N.; Johnston, K. L.; Knight, S.; Kuhnert, N. Hierarchical scheme for LC-MSn identification of chlorogenic acids. J. Agric. Food Chem. 2003, 51, pp. 2900–2911.

(36) Laleh, G. H.; Frydoonfar, H.; Heidary, R.; Jameei, R.; Zare, S. The effect of light, temperature, pH and species on stability of anthocyanin pigments in four berberis species. Pak. J. Nutr. 2006, 5, pp. 90–92.

(37) Davis, B. D.; Brodbelt, J. S. Regioselectivity of human UDP-glucuronosyl-transferase 1A1 in the synthesis of flavonoid glucuronides determined by metal complexation and tandem mass spectrometry. J. Am. Soc. Mass Spectrom. 2008, 19, pp. 246–256.

(38) Gamage, N.; Barnett, A.; Hempel, N.; Duggleby, R. G.; Windmill, K. F.; Martin, J. L.; McManus, M. E. Human sulfotransferases and their role in chemical metabolism. Toxicol. Sci. 2006, 90, pp. 5–22.

(39) Crespy, V.; Nancoz, N.; Oliveira, M.; Hau, J.; Courtet-Compondu, M.-C.; Williamson, G. Glucuronidation of the green tea catechins, (-)-epigallocatechin-3-gallate and (-)-epicatechin-3-gallate, by rat hepatic and intestinal microsomes. Free Radical Res. 2004, 38, pp. 1025–1031.

(40) Feliciano, R. P.; Mecha, E.; Bronze, M. R.; Rodriguez-Mateos, A. Development and validation of a high-throughput micro solid-phase extraction method coupled with ultra-high-performance liquid chromatography-quadrupole time-of-flight mass spectrometry for rapid identification and quantification of phenolic metabolites in human plasma and urine. J. Chromatogr. A. 2016, 1464, pp. 21–31.

(41) de Ferrars, R. M.; Czank, C.; Zhang, Q.; Botting, N. P.; Kroon, P. A.; Cassidy, A.; Kay, C. D. The pharmacokinetics of anthocyanins and their metabolites in humans. Br. J. Pharmacol. 2014, 171, pp. 3268–3282.

Chapter 3

Chemical Hemisynthesis of Sulfated Cyanidin-3-*O*-Glucoside and Cyanidin Metabolites

The metabolism of anthocyanins in humans is still not fully understood, which is partly due to the lack of reference compounds. It is known that sulfation is one way of the complex phase II biotransformation mechanism. Therefore, cyanidin-3-*O*-glucoside and the cyanidin aglycone were chemically converted to their sulfates by reaction with sulfur trioxide-N-triethylamine complex in dimethylformamide. The reaction products were characterized by UHPLC coupled to linear ion trap and IMS-QTOF mass spectrometry. Based on MS data, retention times, and UV-Vis spectra, the compounds could tentatively be assigned to A-, C-, or B-ring sulfates. Analysis of urine samples from two volunteers after ingestion of commercial blackberry nectar demonstrated the presence of two sulfated derivatives of the cyanidin aglycone and one sulfated derivative of the cyanidin-3-*O*-glucoside. It was found that both the A ring and the B ring are sulfated by human enzymes. This study marks an important step toward a better understanding of anthocyanin metabolism.

Keywords: anthocyanins; cyanidin-3-*O*-glucoside; cyanidin; LC-IMS; HRMS; sulfation; metabolites

This chapter has been published:

Straßmann, S.; Passon, M.; Schieber, A. Chemical Hemisynthesis of Sulfated Cyanidin-3-*O*-Glucoside and Cyanidin Metabolites. *Molecules* **2021**, 26, 2146.

1 Introduction

Anthocyanins are water-soluble flavonoids that are widely found in the plant kingdom, for example, in berry fruits. Whereas the profile of some berries such as blueberries is rather complex, other fruits like strawberries and blackberries show a relatively simple composition, with pelargonidin-3-*O*-glucoside and cyanidin-3-*O*-glucoside respectively being the predominant compounds.[1] While it is well known that anthocyanins are poorly bioavailable, a large body of evidence suggests that they impart health benefits. Because anthocyanins have been demonstrated to be extensively metabolized, it is quite obvious that the metabolites rather than the parent compounds constitute the bioactive agents.[2] Most research on anthocyanin metabolites has focused on low-molecular phenolic compounds that result from the degradation of the flavonoid aglycone. However, metabolism of the intact anthocyanins and the aglycone may also occur, for example, by sulfation, glucuronidation, and methylation. In contrast to studies on the latter two pathways, the sulfation of anthocyanins has less frequently been reported, which may be due to the fact that sulfated metabolites are formed less often,[3] the lack of reference compounds, and difficulties in sample preparation. However, the availability of standard substances is a prerequisite for the elucidation of the pharmacokinetics of anthocyanins. Significant progress has been made in the synthesis of metabolites by enzymatic routes. For example, sulfated derivatives of cyanidin and cyanidin-3-*O*-glucoside were obtained employing porcine liver enzymes[4] and may be used for reference in databases. Because the yields were poor and enzymatic hemisynthesis is expensive, alternate approaches are needed. Chemical hemisynthesis may be a more suitable tool to generate sulfated metabolites of anthocyanins. However, the sensitivity of anthocyanins to light, pH changes, and high temperature poses a problem,[5] as does the lability of the sulfate groups under acidic conditions and at high temperatures.[6] As a result of these stability issues, synthesis should be carried out with as few steps as possible and under mild conditions. In anthocyanins, two types of hydroxyl groups are present, which are the aromatic OH groups of the anthocyanidin aglycone and the aliphatic OH groups of the sugar moiety. Because to our knowledge, there is no report about sulfations in the human body taking place on the sugar hydroxyl groups of anthocyanins, we aimed at the chemical hemisynthesis of anthocyanins sulfated on the flavonoid moiety. For this purpose, we used cyanidin-3-*O*-glucoside obtained from blackberry juice and the cyanidin aglycone prepared according to a previously described protocol.[4] Subsequently, we conducted a pilot study with two participants, who consumed blackberry juice and provided urine samples for analysis by LC-MS. This, in turn, was compared to the synthesized reference substances. Among other phase II metabolites of cyanidin, sulfated cyanidin-3-*O*-glucoside and sulfated cyanidin were detected. The aim of this work was to generate authentic reference substances for the identification of sulfated metabolites in real human biological samples.

2 Results and discussion

The present study aimed at the chemical hemisynthesis of sulfated cyanidin-3-*O*-glucoside and cyanidin. For this purpose, the anthocyanin was purified from blackberry juice because blackberries have a relatively simple profile of pigments, cyanidin-3-*O*-glucoside being the predominant compound. Subsequently, the aglycone cyanidin was released by acidic hydrolysis. Both compounds reacted with a sulfur trioxide-N-triethylamine complex to yield a mixture of different mono- and disulfates (**Scheme 1**).

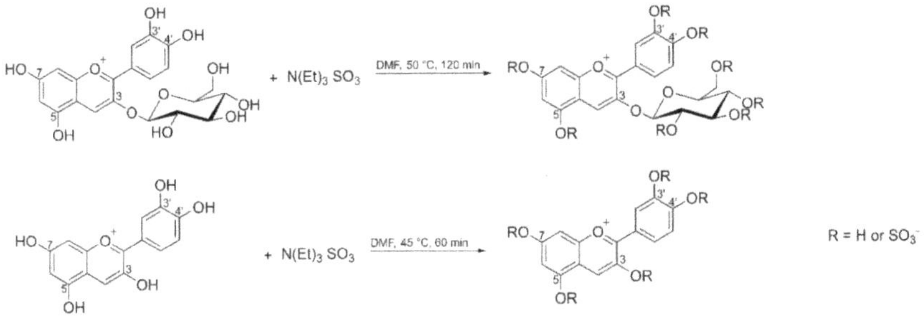

Scheme 1: Synthesis reactions of cyanidin sulfates and cyanidin-3-*O*-glucoside sulfates.

2.1 Cyanidin Sulfates

The extracted ion chromatogram of the $[M]^+$ of monosulfated cyanidin is shown in **Figure 1A**. Five peaks belonging to monosulfates (**1–5**) and at least two peaks (RT 4.5 and 11.3 min) which are in-source fragments of disulfates can be observed. While absolute quantification was not possible due to the lack of references, the relative distribution of the monosulfates could be determined. Thus, compound **1** was formed at 4%, compound **2** at 1%, compound **3** at 3%, compound **4** at 50%, and compound **5** at 42%.

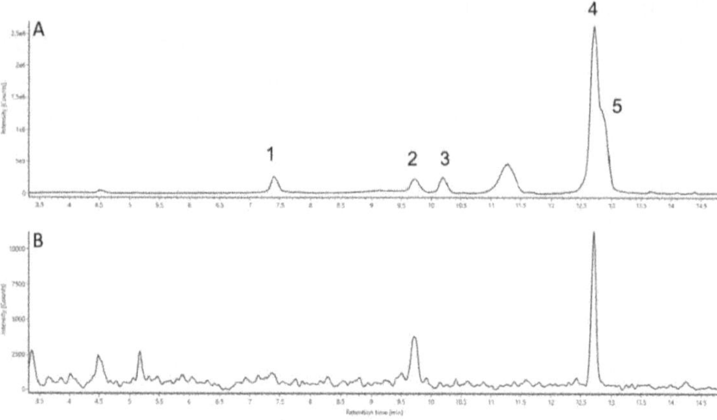

Figure 1: Extracted ion chromatograms of the $[M]^+$, *m/z* 367.0118 (cyanidin monosulfate) after the synthesis reaction (A) and of the 24 h urine sample (B).

Identification was based on fragmentation patterns, accurate masses, CCS (collision cross-section) values, chromatographic behavior, and UV-spectra (**Table 1**). The interpretation of these data allows a tentative assignment of the site of sulfation to the A, B or C ring, which is described as follows.

Table 1: LC-DAD-ESI-LIT and LC-DAD-ESI-IMS-QTOF results of the synthesis reaction of cyanidin sulfates

peak no.	RT^a (min)	$[M]^+$ calculated (*m/z*)	$[M]^+$ measured (*m/z*)	mass error (mDa)	characteristic fragment ions MS^2 (*m/z*) QTOF	characteristic fragment ions MS^2 (*m/z*) LIT	degree of sulfation	CCS value /$[M]^+$ (Å^2)	λ_{max} (nm)	supposed location of sulfate group
-	4.4	446.9686	446.9691	+ 0.5	286.0469, 367.0120	287, 367	2	198.9 ± 0.6	250, 385, 505	n.d.
-	4.8	446.9686	446.9684	− 0.2	287.0550, 367.0117	287, 367	2	196.3 ± 0.1	n.d.	n.d.
-	5.5	446.9686	446.9692	+ 0.6	287.0552, 367.0120	287, 367	2	204.5 ± 0.6	278, 487	n.d.
-	6.5	449.1078	449.1079	+ 0.1	287.0551	287	*	204.6 ± 0.2	278, 516	
1	7.5	367.0118	367.0118	± 0.0	121.0285, 137.0234, 149.0235, 177.0183, 256.9752, 286.0492, 287.0551	137, 177, 257, 287	1	183.6 ± 0.3	267, 410, 505	C ring
-	7.6	446.9686	446.9689	+ 0.3	287.0550, 367.0121	287, 367	2	203.7 ± 0.3	281, 505	n.d.
-	9.3	446.9686	446.9686	± 0.0	286.0473, 367.0123	287, 367	2	202.7 ± 0.1	254, 372, 490	n.d.
2	9.9	367.0118	367.0118	± 0.0	109.0284, 121.0285, 137.0235, 139.0390, 149.0238, 177.0185, 216.9805, 218.9785, 269.0447, 286.0475, 287.0551,	137, 139, 177, 217, 257, 269, 287	1	183.3 ± 0.1	275, 513	A ring
3	10.3	367.0118	367.0120	+ 0.2	121.0286, 137.0236, 139.0392, 149.0239, 177.0184, 219.0265, 256.9752, 287.0552,	137, 177, 257, 287	1	187.6 ± 0.1	281, 517	A ring
-	11.3	446.9686	446.9687	+ 0.1	286.0470, 367.0120	287, 367	2	189.6 ± 0.2	278, 329sh, 428, 497	n.d.
-	11.7	287.0550	287.0552	+ 0.2	109.0286, 121.0286, 137.0236, 149.0236, 269.0448	137, 149, 177, 269	cyanidin	163.4 ± 0.1	277, 443sh, 525	
4	12.6	367.0118	367.0118	± 0.0	216.9807, 268.0370, 286.0471	137, 269, 287	1	182.6 ± 0.1	271, 429, 513	B ring
5	12.8	367.0118	367.0118	± 0.0	268.0370, 269.0445, 286.0472, 287.0553	137, 269, 287	1	180.8 ± 0.2	283, 512	B ring

*Cyanidin-3-*O*-glucoside; n.d. = not determined; a = gradient a

2.1.1. Retention Time

The retention time may be indicative of the site of sulfation and should be taken into consideration for tentative peak assignment. From **Figure 1A** it can be seen that peak **1** elutes very early and remote from all other cyanidin sulfates. The chromatographic behavior of peaks **4** and **5** is very similar, resulting in partial coelution, which suggests a structural similarity. According to previous studies, quercetin sulfated at the 3-OH position elutes earlier than derivatives that bear sulfate groups on the 3'- or 4'-OH positions.[7,8] The elution order found for quercetin monosulfates was 3, 5, 7, 4', and 3'.[9] Taking into account the structural similarity of flavonoids and the identical substitution pattern of the B ring of quercetin and cyanidin, it is reasonable to assume this elution order also for cyanidin monosulfaltes.

2.1.2 UV/Vis Spectra

Consideration of the shifts of the λ_{max} is another useful way of identifying the individual compounds formed in this reaction. Anthocyanins show a distinctive band I in the 450–560 nm region and band II in the 240–280 nm region.[10,11] Sulfation of cyanidin led to a hypsochromic shift in the spectrum of all peaks **1–5**. Peak **1** showed the largest shift in comparison to cyanidin in band I from 525 nm to 505 nm and in band II from 277 to 267 nm. Shifts in the range of 2 to 13 nm were observed for all other peaks. A comparison of the UV/Vis spectra of cyanidin, peak **1** and the two most abundant monosulfated cyanidin derivatives (peaks **4** and **5**) is shown in **Figure 2**. Hypsochromic shifts of the absorption maxima of quercetin through sulfation are already reported in the literature.[9] Here, the 3-OH sulfation caused also the highest shift of 21 nm in band I, whereas conjugation at the 5-, 4'-, and 3'-OH resulted in a small shift; sulfation of the 7-OH position did not affect the λ_{max}. Interestingly, the formation of a new band at approximately 420 nm was observed, which constitutes an even bigger difference between the sulfated cyanidin molecules than the shift of the maxima. Peaks **1** and **4** show this band at 411 nm and 425 nm respectively, whereas sulfation at the other positions either does not affect this band, or the compound concentrations were too low. However, it is possible to determine the ratio of the absorption at 440 nm and the absorption maximum ($\lambda_{440nm}/\lambda_{max}$). This ratio is 70% for peaks **1–3**, 50% for peak **4** and 30% for peak **5**, cyanidin, and cyanidin-3-O-glucoside. This ratio is already described in the literature,[12] where it is reported that the ratio of an anthocyanin 3-O-glucoside is twice as high as that of a 5-O-glucoside. Larger shifts are caused by the disulfates, as can be seen from **Table 1**. Due to the high number of possible combinations and low substance concentrations, no conclusion concerning the conjugation positions can be drawn.

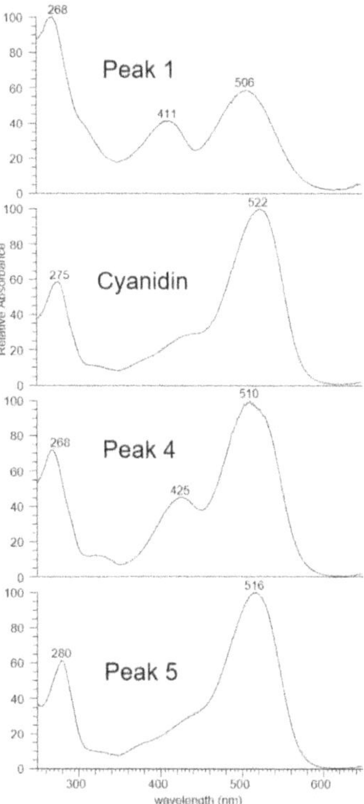

Figure 2: UV/Vis spectra of cyanidin (RT 11.7 minutes) in comparison to three cyanidin monosulfates (RT 7.1, 12.6 and, 12.8 minutes).

2.1.3 Mass Spectrometry

Mass spectrometry is one of the most valuable tools for the characterization of compounds and may even be superior to NMR spectroscopy when only small amounts are available. In the case of flavonoids, the small fragments emerging from the cleavage of the C ring are most valuable for identification purposes. The introduction of a sulfate group leads to a mass increment of 79.9569 u in the spectrum. These characteristic fragments and their corresponding sulfated forms are shown in **Figure 3** and the masses observed via LIT-MS or TOF-MS analyses are listed in **Table 1**. Because the flavylium cation is very stable and the sulfate group represents a good leaving group leading to a neutral loss fragment, only limited fragmentation occurs and the fragments smaller than 286/287 u ($[M - 80]^+$) are of little diagnostic benefit.

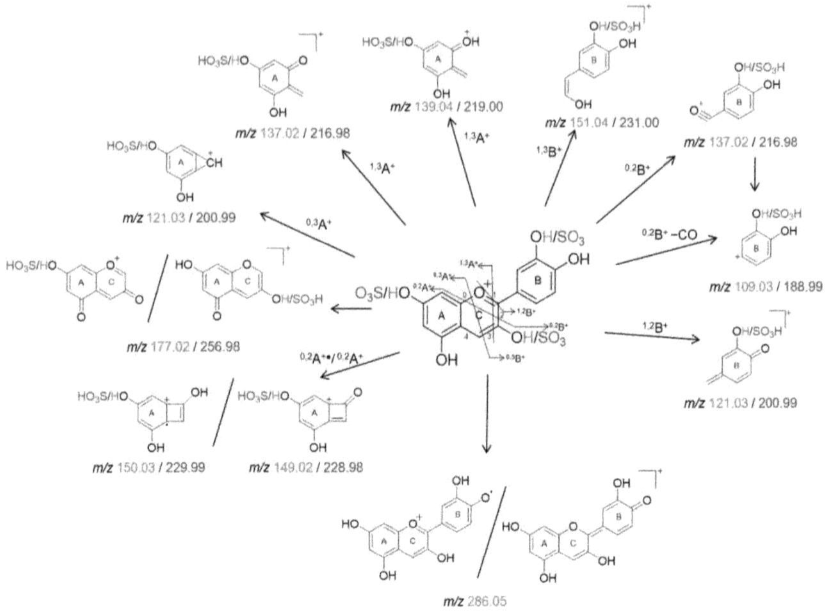

Figure 3: Postulated ESI+ fragmentation scheme for cyanidin (red) and its derivatives (blue) sulfated on the A, B or C ring, based on the proposals of Barnes and Schug[13], González-Manzano et al.[14], and Oliveira et al.[15]. Conjugation positions are arbitrarily and may occur at any OH group.

However, a tentative assignment of the individual peaks to a conjugation position on the A-, B- or C ring is possible. Peak **1** shows numerous small fragments. The fragment at *m/z* 149.02 originates from the cleavage of the A ring ($^{0,2}A^+$). Since it is not present in sulfated form, sulfation on the A ring can be excluded. Furthermore, fragments at *m/z* 121.03 ($^{0,3}A^+$ or $^{1,2}B^+$) and 137.02 ($^{1,3}A^+$ or $^{0,2}B^+$) were obtained, which may be both A- and B ring fragments. Because no fragment corresponding to the sulfated form was observed, it can be concluded that sulfation took place on the C ring. This assumption is confirmed by the simultaneous presence of a fragment at *m/z* 177.02 ($[M - C_6H_5O_2]^+$) and its sulfated form at *m/z* 256.98. The spectra of peaks **2** and **3** look very similar, in here fragments at *m/z* 137.02 ($^{1,3}A^+$ or $^{0,2}B^+$), 139.04 ($^{1,3}A^+$), and the respective sulfated forms at *m/z* 216.98 and 219.00 are detectable. In addition, fragments at *m/z* 177.02 ($[M - C_6H_5O_2]^+$) and 256.98 are present. Furthermore, the fragment at *m/z* 109.03 ($^{0,2}B^+ - CO$) but not its sulfated version can be found. These combinations lead to the conclusion that the A ring carries the sulfate group in these two compounds. Peaks **4** and **5** do not show characteristic fragments, but also look very similar. However, applying the exclusion principle and considering their very similar retention times it can be concluded that these compounds correspond to cyanidin sulfated at the 3'-OH and 4'-OH groups of the B ring, respectively. All peaks show a fragment at *m/z* 286.05, but with different intensities.

Interestingly, the ratio of the fragment at *m/z* 286.05 to 287.02 is relatively constant and is 0.04 for peak **1**, 0.07 for peak **2**, and 0.01 for peak **3**. In peak **4**, this ratio reverses to 17 and the fragment at *m/z* 287.02 is hardly detectable. The last peak (**5**) shows both fragments, in some analyses with a ratio of 0.07 and in some with a ratio of 17. A possible explanation is that the fragment at *m/z* 286.05 belongs to one of the structures shown in **Figure 3** ($[M - 80]^{+}$ or $[M - 80]^{\bullet +}$) and may be related to the radical scavenging properties of flavonoids.[16] The *o*-dihydroxy structure in the B ring confers high stability to a radical form and participates in electron delocalization.[17] It should be mentioned that some fragments were found only in the mass spectra recorded with the ion trap and some only in the spectra recorded with the TOF-MS. These differences may be due to the different collision gases used. The fragments are listed in **Table 1** and therefore the differences are not explicitly mentioned in the descriptions of the spectra. Another useful tool to distinguish compounds is ion mobility spectrometry and the resulting CCS values. If the measurements were taken on the same day, the values are very constant. In the case of the cyanidin sulfates, the CCS values are in the range of 180.8 $Å^2$ for a probable B ring sulfate and 187.6 $Å^2$ for a probable A ring sulfate. Thus, the distinction of the individual components can be accomplished based on these CCS values. Founded on the data discussed above, the elution sequence of the monosulfated compounds is C ring sulfate < A ring sulfate < B ring sulfate.

2.2 Cyanidin-3-*O*-Glucoside Sulfates

In contrast to the cyanidin aglycone, there is no free OH group in the C ring that is available for sulfation. Therefore, conjugation of cyanidin-3-O-glucoside can take place only on the A- and the B rings. **Figure 4** shows extracted ion chromatograms at *m/z* 529.06 and 609.02, corresponding to the masses of mono- and disulfated cyanidin-3-*O*-glucoside.

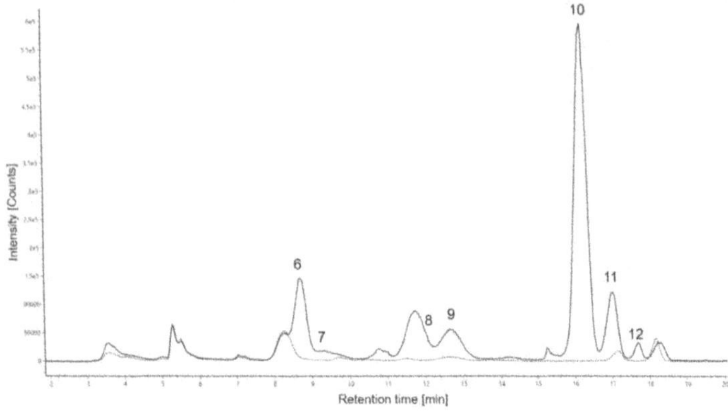

Figure 4: Extracted ion chromatograms of the [M]$^+$, *m/z* 529.0646 (black line) and 609.0215 (dashed line) (cyanidin-3-*O*-glucoside mono- and disulfate) after the synthesis reaction.

The yields determined at 280 nm were 4.4% for compound **6**, 3.0% for compound **8**, 3.8% for compound **9**, and 26.9% for compound **10**. All other monosulfated compounds were obtained at yields under 1%. **Table 2** illustrates the data used for identification purposes, that is, retention times, accurate masses, CCS values, *m/z* of the most abundant fragments, relative yield, and absorption maxima.

Table 2: LC-DAD-ESI-LIT and LC-DAD-ESI-IMS-QTOF results of the synthesis reaction of cyanidin-3-*O*-glucoside sulfates

peak no.	RT^b (min)	[M]$^+$ calculated (*m/z*)	[M]$^+$ measured (*m/z*)	mass error (mDa)	fragment ions MS^2 (*m/z*)	degree of sulfation	CCS value [M]$^+$ (Å^2)	λ_{max}
-	5.3	609.0215	609.0221	+ 0.6	529 (3.5), 447 (100), 367 (4.2)	2	240.8 ± 0.1	277, 505
-	5.5	609.0215	609.0219	+ 0.4	529 (5.1), 447 (100), 367 (5.4)	2	240.6 ± 0.1	278, 505
-	7.0	609.0215	609.0225	+ 1.0	529 (100), 447 (88.1), 367 (52.3), 287 (4.0)	2	237.5 ± 0.5	278
-	8.3	609.0215	609.0224	+ 0.9	529 (70.9), 367 (100), 287 (14.5)	2	232.3 ± 0.1	278, 500
6	8.8	529.0647	529.0650	+ 0.3	449 (2.8), 367 (100), 287 (5.3)	1	224.7 ± 0.1	278, 509
7	9.3	529.0647	529.0649	+ 0.2	449 (96.5), 367 (3.0), 287 (100)	1	217.1 ± 0.1	277, 504
-	9.8	609.0215	609.0223	+ 0.8	529 (100), 447 (2.3), 367 (18.7), 287 (3.3)	2	239.9 ± 0.2	278, 509
-	11.5	609.0215	609.0222	+ 0.7	529 (57.4), 447 (1.9), 367 (100), 287 (27.4)	2	236.2 ± 0.1	278, 492
8	12.0	529.0647	529.0652	+ 0.5	449 (7.5), 367 (100), 287 (4.5)	1	219.2 ± 0.1	278, 511
9	12.8	529.0647	529.0650	+ 0.3	449 (100), 367 (80.2), 287 (92.6)	1	216.2 ± 2.4	277, 407, 493
-	15.2	449.1078	449.1079	+ 0.1	287 (100)	*	204.6 ± 0.2	279, 516
10	16.4	529.0647	529.0651	+ 0.4	449 (100), 287 (99.6)	1	215.7 ± 0.2	279, 516
11	17.1	529.0647	529.0651	+ 0.4	449 (100), 287 (35.2)	1	218.2 ± 0.1	280, 516
-	17.2	609.0215	609.0224	+ 0.9	529 (100), 449 (2.3), 367 (1.8), 287 (19.3)	2	224.7 ± 0.2	280, 516
12	17.7	529.0647	529.0648	+ 0.1	449 (17.2), 287 (100)	1	214.9 ± 0.1	280, 516
-	18.2	609.0215	609.0220	+ 0.5	529 (100), 449 (21.8), 367 (1.1), 287 (3.9)	2	224.4 ± 0.1	280, 517

*Cyanidin-3-*O*-glucoside; b = gradient b

2.2.1. Retention Time

Four compounds (**6**, **7**, **8**, **9**) at *m/z* 529.06 elute earlier and three (**10**, **11**, **12**) later than cyanidin-3-*O*-glucoside. From this elution behavior it can be concluded that these two clusters differ in the site of conjugation. Furthermore, the first cluster can be divided into two groups. Peaks **6** and **7** elute at similar retention times as well as peaks **8** and **9**. Considering the elution order of the sulfated cyanidin aglycone and assuming that the sugar moiety does not influence the elution behavior, the two earlier eluting peaks can be allocated to A ring sulfates and the other two to B ring sulfates.

2.2.2 UV Spectra

UV spectral data obtained for sulfation of cyanidin-3-*O*-glucoside were generally similar to those described for the aglycone. Because of the low concentrations, however, intensities were less pronounced. The absorptions at approximately 280 nm remained largely unchanged. Sulfation of compound **9** led to the largest shift of the λ_{max}, that is, from 516 nm for cyanidin-3-*O*-glucoside to 493 nm. The additional band at about 420 nm already described for the cyanidin sulfates (peak **1** and **4**) was observed at 407 nm. Based on this data, the above-mentioned peak assignment is confirmed.

2.2.3 Mass Spectrometry

The mass spectrometric data (**Table 2**) allowed the distinction of the above-mentioned two clusters of peaks. Compounds eluting earlier than 15 min showed a neutral loss of 162 u caused by the loss of the glucose moiety. These carry the sulfate group at the aglycone. In contrast, for molecules eluting later than 15 min, a neutral loss of 242 u was observed, which is indicative of a sulfate group at the sugar moiety. Unfortunately, an unambiguous assignment is not possible to a distinct position by means of mass spectrometry because the sulfate group is an excellent neutral loss fragment. However, the intensities of the resulting fragment signals (*m/z* 287.06 $[M-80-162]^+$, 367.01 $[M-162]^+$, and 449.11, $[M-80]^+$) differ for each peak (**Table 2**). This behavior can be exploited, in addition to the retention time, for the identification of metabolites in physiological samples when their isolation is not possible, and references are not available. Because the four peaks eluting early (compounds **6**, **7**, **8** and **9**) could be allocated to a sulfated aglycone, it is evident that all hydroxyl groups are sulfated, yet some of them seem to be preferred. Whereas compound **6** was obtained at a yield of 4.5%, compound **7** was produced to such a small extent that no yield could be determined. This preference might be attributed to the differences in the pK_a value of the hydroxyl groups, which is the lowest for the 5-OH group.[18] In the case of the sugar moiety, steric reasons may be the explanation for differences in the site of sulfation.[19] Furthermore, higher temperatures during the reaction lead to an increased formation of disulfates, which is described in the literature[19] and which was also observed in this work. Interestingly, we observed decreasing CCS values with increasing retention times within the monosulfates or disulfates, respectively. It may be concluded that smaller molecules, whose size can be derived from the shorter drift times, interact more with the column material, which results in longer retention times. This observation was also made with the metabolites that resulted from cyanidin and cyanidin-3-*O*-glucoside through liver enzymes.[4]

2.3 Challenges

Some difficulties with the sulfation of small molecules are already described in the literature.[6] As inorganic salts may pose a problem, particularly in smaller approaches, they may cause inconsistencies concerning the purification and thus the yield. While the separation of quercetin sulfates by preparative HPLC has already been reported to pose a problem,[20] it may be even more difficult in the case of anthocyanins because of their pH-dependent structure. Additionally, the instability of the sulfate groups under acidic conditions and at high temperatures are pointed out in the literature[6]. The synthesis of resveratrol sulfates has even been referred to as a nightmare that can be handled only by selective protection of the OH groups.[21] Protective groups, which are often used for flavonoid derivatization, should be avoided for anthocyanins because their introduction and removal are often associated with unfavorable conditions. In contrast, the method applied in this study allows the production of sulfated derivatives without using protective groups. Another issue is the solvolysis of phenolic sulfates, which has been reported for 3- and 4-*O*-resveratrol sulfates during evaporation to dryness, probably caused by the presence of sulfuric acid as a reaction byproduct.[22] These and other challenges are the reasons why anthocyanins rarely appear in the long history of sulfation of phenolic substances.[23] Furthermore, with respect to the intensities, it should also be considered that the anthocyanin monosulfates may be converted to disulfates. Therefore, fewer monosulfates should be present at higher temperatures and longer reaction times. Although LC-MS analysis is extremely useful for metabolite characterization, challenges still exist in the assignment of the peaks to the right number of sulfates. The retention times of some of the monosulfates are close to the disulfates and the disulfates produce monosulfates as in source fragments.

2.4 Human Pilot Study

In a first pilot study comprising only two individuals, we investigated urine samples for the presence of anthocyanin phase II metabolites after ingestion of a commercial blackberry nectar sample. The qualitative profile of the investigated metabolites in the two urine samples was comparable. Their analytical data such as the accurate mass, the main fragment, and the CCS values are given in **Table 3**. Among the metabolites detected, two sulfated cyanidin derivatives and one sulfated cyanidin-3-*O*-glucoside were found. The latter elutes after 13.1 min (gradient b), which corresponds to the retention time of the reference sulfate **9**. Their CCS values are also in the same range. However, it should be noted that while the deviations observed within a measurement are remarkably small, larger differences between different measurements may occur. For example, day-to-day variations in the CCS value for cyanidin of 162.5 ± 0.1 Å² and on another day 165.5 ± 0.1 Å² were observed, which results in an overall deviation of ± 2.1 Å². Thus, the deviation of the CCS value of the synthesized sulfate **9** (216.2 ± 2.4 Å²) from the sulfate found in urine (209.4 ± 0.6 Å²) with nearly the same retention time can be justified.

Table 3: LC-DAD-ESI-IMS-QTOF results of the analysis of urine collected for 24 h after the consumption of blackberry juice, either gradient a or b

suspected compound	RT (min)	[M]⁺ calculated (*m/z*)	[M]⁺ measured (*m/z*)	mass error (mDa)	fragment MS² (*m/z*)	CCS value (Å²)
			cyanidin			
unconjugated*	11.7ᵃ	287.0550	287.0549	− 0.1		162.5 ± 0.1
glucuronidated	8.3ᵃ	463.0877	463.0871	− 0.6	287	200.1 ± 0.9
	9.6ᵃ	463.0877	463.0874	− 0.3	287	204.7 ± 1.0
glucuronidated and methylated	8.8ᵃ	477.1033	477.1029	− 0.4	301	209.4 ± 2.4
	9.1ᵃ	477.1033	477.1030	− 0.3	301	207.6 ± 0.7
	9.6ᵃ	477.1033	477.1028	− 0.5	301	208.2 ± 0.7
	10.2ᵃ	477.1033	477.1029	− 0.4	301	207.6 ± 0.4
sulfated	9.8ᵃ	367.0118	367.0118	± 0.0	287	180.0 ± 1.0
	12.7ᵃ	367.0118	367.0118	± 0.0	287	180.6 ± 0.5
			cyanidin-3-*O*-glucoside			
unconjugated*	15.0ᵇ	449.1078	449.1081	+ 0.3	287	203.3 ± 2.5
sulfated	13.1ᵇ	529.0647	529.0648	+ 0.1	287	209.4 ± 0.6

a = gradient a; b = gradient b; *Not found in the samples, listed only for comparison

The earliest peak corresponding to cyanidin sulfates that were detected in the urine samples elutes after 9.8 min (gradient a), which matches the retention time of peak **2**. The CCS value of this compound is 180.0 ± 1.0 Å² and hence approximately 2 Å² below that of the reference compound, which can be explained by the fact that measurements were not taken on the same day. Taking into account these deviations, the compound eluting after 12.7 min can be assigned to peak **4** of the synthesized sulfates. The CCS value also of this peak **4** is approximately 2 Å² higher than that of the substance detected in the urine. Based on the mass spectrometric data, the compound eluting at 9.8 min can be assigned to an A ring sulfate, whereas the compound eluting after 12.7 min is sulfated at the B ring. To the best of our knowledge, this is the first work that reports a flavonoid metabolized to an A ring sulfate. Studies conducted on quercetin demonstrated the formation of quercetin-3'-*O*-sulfate in rat liver,[7] which is a metabolite also present in humans.[8] From these findings, it may be deduced that compound **4** corresponds to cyanidin-3'-*O*-sulfate. As the 3'-*O*-sulfate was also obtained in the highest yields by other chemical syntheses,[7] it may be concluded that the most intense peak in the synthesis of cyanidin sulfates is the 3'-*O* sulfate. These results would confirm that the human body produces mainly 3'-*O*-sulfates of flavonoids.[8]

3 Materials and methods

3.1 Chemicals

LC-MS-grade water, acetonitrile, formic acid, and ammonium hydroxide were purchased from ChemSolute (Renningen, Germany). Sulfur trioxide-N-triethylamine complex, leucine enkephalin acetate salt hydrate, and pyridine were obtained from Sigma-Aldrich (St. Louis, MO, USA). N,N-Dimethylformamide (DMF) was from Fisher Scientific (Waltham, MA, USA). Methanol, ascorbic acid, and hydrochloric acid (HCl) were supplied by VWR (Darmstadt, Germany). Trifluoroacetic acid (TFA) was purchased from Alfa Aesar (Ward Hill, MA, USA). An anthocyanin extract containing primarily cyanidin-3-*O*-glucoside (**Scheme 2**) was obtained as described before.[4] To obtain the cyanidin aglycone, the lyophilized extract was treated with concentrated HCl for 90 min at 90 °C.

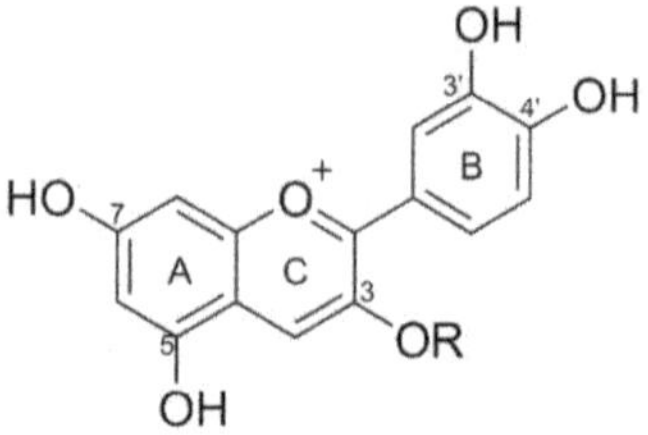

OH
3' 4' OH
B
HO 7 O+
A C 3
5 OR
OH

R = H = cyanidin
R = glucose = cyanidin-3-*O*-glucoside

Scheme 2: Structure of cyanidin and cyanidin-3-*O*-glucoside

3.2 Hemisynthesis of Sulfated Metabolites

The approach is based on the method described by Jones et al.[7] and the equations in **Scheme 1** graphically illustrate the two reactions of cyanidin and cyanidin-3-*O*-glucoside, respectively. Cyanidin-3-*O*-glucoside (7 µmol) was dissolved in approximately 20 mL dry pyridine to remove any water that may be present. Pyridine was evaporated under reduced pressure and the entire procedure was repeated. The dried compound was dissolved in 1 mL DMF. The cyanidin aglycone (7 µmol) was dissolved directly in 1 mL DMF. This mixture was then transferred to an Eppendorf tube and allowed to react with a 10-fold excess of sulfur trioxide-N-triethylamine complex in a thermomixer (Eppendorf, Hamburg, Germany). A temperature of 50 °C and a reaction time of 120 min were found to be optimal for the hemisynthesis of monosulfated cyanidin-3-*O*-glucoside; for the cyanidin aglycone, optimum conditions were 45 °C and 60 min, respectively. These conditions allowed both maximizing the concentration of the monosulfates and minimizing the formation of multiple sulfated conjugates.

After 5-fold dilution with water and acetonitrile containing 0.1% formic acid, and microfiltration (regenerated cellulose, 0.20 µm), the samples were analysed using LC-LIT-MS^n and LC-IMS-QTOF-MS. The polarity of non-charged flavonoids is increased by sulfation in a way that they precipitate in the used solvent, which is mostly dioxane.[7] In the case of anthocyanins, the polarity of the resulting substances is similar to the parent molecules, so that they do not precipitate. Therefore, solid-phase extraction with a weak anion exchanger was applied to separate the cyanidin-3-*O*-glucoside sulfates from the starting material to take advantage of the different local charge distribution. For this purpose, OASIS WAX cartridges (6 cc Vac cartridge, 500 mg sorbent per cartridge, 60 µm particle size; Waters, Milford, MA, USA) were washed with methanol and water. After dilution of the samples with 4 mL water, aliquots of 2 mL were applied to the ion exchanger; subsequently, the material was washed again with water containing 2% of formic acid. Most of the unreacted cyanidin-3-*O*-glucoside was removed using 100% methanol in the first elution step. The sulfated target compounds were then eluted with 5% ammonia in methanol, 3% HCl in water/methanol 50:50 (*v/v*), and 20% trifluoracetic acid in water/methanol 50:50 (*v/v*) until the eluates were colorless (approx. 3–5 mL). Eluates were evaporated under a constant nitrogen stream to dryness and redissolved in 250 µL of acetonitrile/water/formic acid 20:79.9:0.1 (*v/v/v*).

3.3 LC-MS-Analysis

UHPLC analysis of the reaction products was performed on an Acquity UPLC Iclass system (Waters, Milford, MA, USA) consisting of a binary pump, a sample manager cooled at 10 °C, a column oven set at 40 °C, and a diode array detector scanning from 250 to 650 nm. An Acquity HSS-T3 RP18 column (150 mm x 2.1 mm; 1.8 µm particle size) combined with a pre-column (Acquity UPLC HSS T3 VanGuard, 100 Å, 2.1 mm x 5 mm, 1.8 µm), both from Waters (Milford, MA, USA) was used for separation with water (A) and acetonitrile (B) as eluents, both acidified with 0.1% (*v* + *v*) formic acid. The flow rate was set at 0.4 mL/min. Since the reaction products differ considerably in their polarity, two different gradients had to be applied to achieve the best separation possible. Sulfated reaction products of the cyanidin aglycone were analyzed using a linear gradient from 8% B to 10% B for 5 min, then to 17% B for 6 min, to 30% B for 4 min, and to 50% B for 8 min (gradient a). Sulfates of cyanidin-3-*O*-glucoside were analyzed using a gradient whose solvent composition changed from 1 to 10% B in 13 min, then to 17% B in 6 min, and then to 30% B in 4 min (gradient b). The injection volume was 5 µL. For MS analysis, the UHPLC was coupled with an LTQ-XL ion trap mass spectrometer (Thermo Scientific, Inc., Waltham, MA, USA) equipped with an electrospray interface operating in the positive ion mode. Ion mass spectra were recorded in the range of *m/z* 245–1000. The capillary was set at 325 °C with a spray voltage of 16 V for cyanidin and 30 V for cyanidin-3-*O*-glucoside. The source voltage was maintained at 4 kV at a current of 100 µA and the tube lens was adjusted to 55 V for cyanidin and 75 V for cyanidin-3-*O*-glucoside. Nitrogen was used as sheath, auxiliary and sweep gas at a flow of 70, 10 and 1 arb, respectively. Three consecutive scans were conducted with helium as collision gas: a full mass scan, an MS^2 scan of the most abundant ion of the first scan using normalized collision energy (CE) of 20%, and an MS^3 analysis of the most abundant ion in the MS^2 experiment with a CE of 35%. In addition, multiple reaction monitoring (MRM) measurements were performed and the masses resulting from typical losses of one $[M - 80]^+$ or two $[M - 80 - 80]^+$ sulfate groups or glucose $[M - 162]^+$ were scanned. Data were acquired and processed using Xcalibur 2.2SP1.48 (Thermo Scientific, Inc., Waltham, MA, USA). For ion mobility spectrometry measurements, the UHPLC was connected to a Vion IMS QTOF mass spectrometer (Waters, Milford, MA, USA) operating in the positive ion mode. The capillary voltage was 0.5 kV, the source temperature was 120 °C, the cone voltage was 40 V, the desolvation gas temperature was 550 °C, and the desolvation gas flow was 1200 L/h. The measurements were conducted with automatic lock correction every 5 min with leucine enkephalin as the lock mass at a concentration of 100 pg/µL. Nitrogen was used as the drift and collision gas and the MS mode was high definition with a low collision energy of 6 eV and a high collision energy ramp of 20–40 eV. Data were acquired and processed using UNIFI v1.9.2.045 (Waters, Milford, MA, USA).

3.4 Study Design and Analysis of Urine

As a proof of concept, the urine of two people was collected over 24 h after consumption of 400 mL commercial blackberry nectar. This volume contained about 100 mg cyanidin-3-*O*-glucoside. Additionally, one sample of urine was taken prior to consumption. The diet was anthocyanin-free in the week before the study and polyphenol-free on the day of the study. According to the procedure of Feliciano et al.[24], ascorbic acid was added to the urine containers (5.63 g/3 L container). Aliquots were adjusted to pH 2.5 with formic acid for stabilization and measured directly to avoid precipitation of the desired metabolites.[25] Direct measurement was required because we had observed that after one month of storage of the samples at −80 °C, the peak intensities were significantly lower.

3.5 Sample Preparation

Samples were prepared according to a previously published protocol.[26] Briefly, 5 mL of a urine sample was evaporated at 40 °C under nitrogen flow to a final volume of 500–700 µL. After microfiltration (regenerated cellulose, 0.20 µm), the samples were analysed using LC-IMS-QTOF-MS and LC-LIT-MS.

4 Conclusions

Our knowledge about the fate of anthocyanins in the human body is still incomplete. It is known that after absorption various enzymes transfer glucuronic acid and/or a sulfate group as well as a methyl group to the aglycone or glycoside. Additionally, cleavage of the glycoside and other reactions are possible. From the results of the human pilot study, it is evident that the metabolite species mentioned above can be found in various forms in urine. It has been possible to chemically synthesize sulfates of cyanidin and cyanidin-3-O-glucoside and to use these as reference substances. Thus, two different cyanidin sulfates and one cyanidin-3-O-glucoside sulfate could be identified. In addition, two glucuronides and four combinations of glucuronidated and methylated derivatives of cyanidin were detected in human urine. Despite the difficulties associated with the small quantities from chemical synthesis, the structural elucidation of the sulfated anthocyanins is an advance in the complete characterization of their metabolic profile.

Author Contributions: Conceptualization, S.S., M.P. and A.S.; methodology, S.S.; validation, S.S.; investigation, S.S.; interpretation, S.S., M.P. and A.S.; resources, A.S.; writing—original draft preparation, S.S.; writing—review and editing, M.P. and A.S.; supervision, M.P. and A.S.; project administration, M.P. All authors have read and agreed to the published version of the manuscript.

Funding: This research received no external funding.

Institutional Review Board Statement: The study was conducted according to the guidelines of the Declaration of Helsinki, and approved by the ethics commission of the Medical Faculty of the University of Bonn (project identification code 019/17).

Informed Consent Statement: Informed consent was obtained from all subjects involved in the study.

Data Availability Statement: Data is contained within the article.

Conflicts of Interest: The authors declare no conflict of interest.

Sample Availability: Samples of the compounds are not available from the authors.

5 References

(1) Horbowicz, M.; Kosson, R.; Grzesiuk, A.; Dębski, H. Anthocyanins of Fruits and Vegetables – Their Occurrence, Analysis and Role in Human Nutrition. Veg. Crops Res. Bull. 2008, 68, 5–22, doi:10.2478/v10032-008-0001-8.

(2) Kay, C.D.; Pereira-Caro, G.; Ludwig, I.A.; Clifford, M.N.; Crozier, A. Anthocyanins and Flavanones Are More Bioavailable than Previously Perceived: A Review of Recent Evidence. Annu. Rev. Food Sci. Technol. 2017, 8, 155–180, doi:10.1146/annurev-food-030216-025636.

(3) Felgines, C.; Talavera, S.; Texier, O.; Gil-Izquierdo, A.; Lamaison, J.-L.; Remesy, C. Blackberry anthocyanins are mainly re-covered from urine as methylated and glucuronidated conjugates in humans. J. Agric. Food Chem. 2005, 53, 7721–7727, doi:10.1021/jf051092k.

(4) Schmitt, S.; Tratzka, S.; Schieber, A.; Passon, M. Hemisynthesis of Anthocyanin Phase II Metabolites by Porcine Liver Enzymes. J. Agric. Food Chem. 2019, 67, 6177–6189, doi:10.1021/acs.jafc.9b01315.

(5) Fossen, T.; Cabrita, L.; Andersen, O.M. Colour and stability of pure anthocyanins influenced by pH including the alkaline region. Food Chem. 1998, 63, 435–440, doi:10.1016/S0308-8146(98)00065-X.

(6) Al-Horani, R.A.; Desai, U.R. Chemical Sulfation of Small Molecules – Advances and Challenges. Tetrahedron 2010, 66, 2907–2918, doi:10.1016/j.tet.2010.02.015.

(7) Jones, D.J.L.; Jukes-Jones, R.; Verschoyle, R.D.; Farmer, P.B.; Gescher, A. A synthetic approach to the generation of quercetin sulfates and the detection of quercetin 3'-O-sulfate as a urinary metabolite in the rat. Bioorg. Med. Chem. 2005, 13, 6727–6731, doi:10.1016/j.bmc.2005.07.021.

(8) Day, A.J.; Mellon, F.; Barron, D.; Sarrazin, G.; Morgan, M.R.; Williamson, G. Human metabolism of dietary flavonoids: iden-tification of plasma metabolites of quercetin. Free Radical Res. 2001, 35, 941–952, doi:10.1080/10715760100301441.

(9) Dueñas, M.; González-Manzano, S.; Surco-Laos, F.; González-Paramas, A.; Santos-Buelga, C. Characterization of sulfated quercetin and epicatechin metabolites. J. Agric. Food Chem. 2012, 60, 3592–3598, doi:10.1021/jf2050203.

(10) Anouar, E.H.; Gierschner, J.; Duroux, J.-L.; Trouillas, P. UV/Visible spectra of natural polyphenols: A time-dependent density functional theory study. Food Chem. 2012, 131, 79–89, doi:10.1016/j.foodchem.2011.08.034.

(11) Welch, C.R.; Wu, Q.; Simon, J.E. Recent Advances in Anthocyanin Analysis and Characterization. Curr. Anal. Chem. 2008, 4, 75–101, doi:10.2174/157341108784587795.

(12) Giusti, M.M.; Wrolstad, R.E. Characterization and Measurement of Anthocyanins by UV-Visible Spectroscopy. Curr. Protoc. Food anal. Chem. 2001, F1.2.1-F1.2.13.

(13) Barnes, J.S.; Schug, K.A. Structural characterization of cyanidin-3,5-diglucoside and pelargonidin-3,5-diglucoside anthocya-nins: Multi-dimensional fragmentation pathways using high performance liquid chromatography-electrospray ionization-ion trap-time of flight mass spectrometry. Int. J. Mass Spectrom. 2011, 308, 71–80, doi:10.1016/j.ijms.2011.07.026.

(14) González-Manzano, S.; González-Paramás, A.; Santos-Buelga, C.; Dueñas, M. Preparation and characterization of catechin sulfates, glucuronides, and methylethers with metabolic interest. J. Agric. Food Chem. 2009, 57, 1231–1238, doi:10.1021/jf803140h.

(15) Oliveira, M.C.; Esperança, P.; Almoster Ferreira, M.A. Characterisation of anthocyanidins by electrospray 78emethylat and collision-induced dissociation tandem mass spectrometry. Rapid Commun. Mass Spectrom. 2001, 15, 1525–1532, doi:10.1002/rcm.400.

(16) Hvattum, E.; Ekeberg, D. Study of the collision-induced radical cleavage of flavonoid glycosides using negative electrospray ionization tandem quadrupole mass spectrometry. J. Mass Spectrom. 2003, 38, 43–49, doi:10.1002/jms.398.

(17) Rice-Evans, C.A.; Miller, N.J.; Paganga, G. Structure-antioxidant activity relationships of flavonoids and phenolic acids. Free Radical Biol. Med. 1996, 20, 933–956, doi:10.1016/0891-5849(95)02227-9.

(18) Dangles, O.; Fenger, J.-A. The Chemical Reactivity of Anthocyanins and Its Consequences in Food Science and Nutrition. Molecules 2018, 23, 1970, doi:10.3390/molecules23081970.

(19) Turvey, J.R.; Williams, T.P. Sulphates of 79emethylated7979s and derivatives. Part V. Products of sulphation of galactose and glucose. J. Chem. Soc. 1963, 2242–2246.

(20) Needs, P.W.; Kroon, P.A. Convenient syntheses of metabolically important quercetin glucuronides and sulfates. Tetrahedron 2006, 62, 6862–6868, doi:10.1016/j.tet.2006.04.102.

(21) Mattarei, A.; Biasutto, L.; Romio, M.; Zoratti, M.; Paradisi, C. Synthesis of resveratrol sulfates: turning a nightmare into a dream. Tetrahedron 2015, 71, 3100–3106, doi:10.1016/j.tet.2014.09.063.

(22) Iwuchukwu, O.F.; Sharan, S.; Canney, D.J.; Nagar, S. Analytical method development for synthesized conjugated metabolites of trans-resveratrol, and application to pharmacokinetic studies. J. Pharm. Biomed. Anal. 2012, 63, 1–8, doi:10.1016/j.jpba.2011.12.006.

(23) Correia-da-Silva, M.; Sousa, E.; Pinto, M.M.M. Emerging sulfated flavonoids and other polyphenols as drugs: nature as an inspiration. Med. Res. Rev. 2014, 34, 223–279, doi:10.1002/med.21282.

(24) Feliciano, R.P.; Boeres, A.; Massacessi, L.; Istas, G.; Ventura, M.R.; Nunes Dos Santos, C.; Heiss, C.; Rodriguez-Mateos, A. Identification and quantification of novel cranberry-derived plasma and urinary (poly)phenols. Arch. Biochem. Biophys. 2016, 599, 31–41, doi:10.1016/j.abb.2016.01.014.

(25) Felgines, C.; Talavéra, S.; Gonthier, M.-P.; Texier, O.; Scalbert, A.; Lamaison, J.-L.; Rémésy, C. Strawberry anthocyanins are recovered in urine as glucuro- and sulfoconjugates in humans. J. Nutr. 2003, 133, 1296–1301.

(26) Kaiser, M.; Müller-Ehl, L.; Passon, M.; Schieber, A. Development and Validation of Methods for the Determination of An-thocyanins in Physiological Fluids via UHPLC-MSn. Molecules 2020, 25, 518, doi:10.3390/molecules25030518.

Chapter 4

Methylation of Cyanidin-3-*O*-Glucoside with Dimethyl Carbonate

The approach presented in this study is the first for the hemisynthesis of methylated anthocyanins. It was possible to obtain cyanidin-3-*O*-glucoside derivatives with different degrees of methylation. Cautious identification of 4'-, 5-, and 7-OH monomethylated derivatives was also accomplished. The methylation agent used was the "green chemical" dimethyl carbonate (DMC), which is characterized by low human and ecological toxicity. The influence of the temperature, reaction time, and amount of the required diazabicyclo[5.4.0]undec-7-en (DBU) catalyst on the formation of the products was examined. Compared to conventional synthesis methods for methylated flavonoids using DMC and DBU, the conditions identified in this study result in a reduction of reaction time, and an important side reaction, so-called carboxymethylation, was minimized by using higher amounts of catalyst.

Keywords: cyanidin-3-*O*-glucoside; cyanidin; anthocyanins; LC-IMS; HRMS; methylation; DMC; phase II metabolites

This chapter has been published:

Straßmann, S.; Brehmer, T.; Passon, M.; Schieber, A. Methylation of Cyanidin-3-*O*-Glucoside with Dimethyl Carbonate. *Molecules* **2021**, 26, 1342.

1 Introduction

Anthocyanins are natural plant colorants and are a component of fruit and vegetables in the human diet.[1] Like other polyphenols, they show antioxidant properties, e.g., against radical oxygen species. As such, they are associated with the prevention of certain diet-related illnesses, such as cardiovascular disease and cancer.[2] The increased interest in anthocyanin-rich nutrition is accompanied by the elucidation of the physiological relationships between intake and metabolism. Metabolites of anthocyanins are inter alia compounds glucuronidated, sulfated, and/or methylated at the aglycone.[3] Knowledge of the structure and quantity of these metabolites after consumption of anthocyanin-rich meals in the human body provides information about the intake and bioavailability and thus, in the long term, about the health significance of anthocyanins in vivo.[4] For such investigations, standard substances are required, and their production gives rise to new research areas.[5] In living organisms, conjugation of flavonoids having a catechol structure with methyl groups is carried out by catechol-*O*-methyl transferase enzymes (COMT) and S-adenosylmethionine (SAM) as a coenzyme.[6] Since cyanidin has this catechol structure only on the B ring, there is no methylation of the A ring. Consequently, methylated derivatives observed so far in vivo carry the methyl group either at the 3'- (peonidin-3-O-glucoside) or at the 4'-hydroxyl group of cyanidin-3-*O*-glucoside (isopeonidin-3-*O*-glucoside).[7] The metabolites themselves can either be excreted directly or be subject to further metabolism. The physiological properties of the metabolites may differ from those of the native starting compounds.[8] Metabolically significant methylated reference substances can be synthesized using different approaches. A promising approach for glucuronidation, sulfation, and methylation is enzymatic hemisynthesis, but costs and efforts are too high to result in usable quantities.[9] Due to the enzyme properties, another drawback of this synthesis route is the lack of the ability to generate derivatives with more than one methyl group, which can be relevant for already genuinely methylated compounds. Thus, chemical synthesis is more promising. This may either mean full synthesis, as performed successfully by Cruz et al.[10], or hemisynthesis. The advantage of hemisynthesis is that only a few steps are required to achieve the goal. Initially, a suitable methylation agent is required. For flavonoids such as quercetin or catechin, successful syntheses with dimethyl sulfate (DMS) or methyl iodide, also known as iodomethane, have been described.[5,11] In addition, mono-, di- and tetramethylated naringenin derivatives were synthesized by using anhydrous acetone and *N,N*-dimethylformamide (DMF), respectively.[12] Since DMS and methyl iodide are toxic and carcinogenic, working with dimethyl carbonate (DMC) is an attractive alternative. For DMC, two reactions are of particular importance: carboxymethylation, also known as methoxycarbonylation, at low temperatures, and methylation at temperatures above 120 °C.[13] Both reactions are shown in **Scheme 1**.

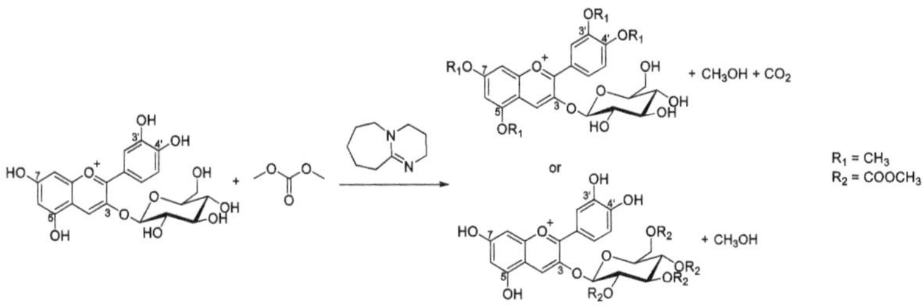

Scheme 1: Possible reactions of cyanidin-3-*O*-glucoside with DMC and DBU.

The crucial difference in carboxymethylation and methylation lies in the mechanism of base-mediated ester hydrolysis of DMC. Successful *O*-methylation (in the following referred to as methylation) depends on a suitable catalyst to prevent carboxymethylation.[14] Methylation and carboxymethylation are also dependent on the type of alcohol.[15] Thus, methylation is preferred on aryl alcohols and carboxymethylation on aliphatic alcohols. Most syntheses of phenols with DMC take place at temperatures above its boiling point, i.e., between 120 and 200 °C.[16,17] Considering the sensitivity of anthocyanins to high temperatures and long reaction times,[18] this is rather critical for their methylation. By using diazabicyclo[5.4.0]undec-7-ene (DBU) as a catalyst, the temperature of the methylation can be reduced below 100 °C,[19] which is nevertheless still too high for the direct synthesis of the cyanidin aglycone. The synthesis of methylated anthocyanins for the exploration of metabolites in physiological samples poses a special challenge, as only aglycone-substituted compounds are relevant. Although the aglycone is an aryl alcohol, the glucose moiety is an aliphatic alcohol. These types differ in the way they react with DMC.

Thus, the aim of the synthesis was to determine the reaction conditions for highly selective *O*-methylation of the aglycone without carboxymethylation of the sugar. For cyanidin-3-*O*-glucoside, these are the derivatives methylated at the B ring.[7] Furthermore, the yield of mono- and demethylated derivatives should be maximized as much as possible, as they are interesting from a metabolic point of view. At the same time, the amount of higher methylated compounds should be minimized.

2 Results and discussion

Using the methods described above, three monomethylated, three dimethylated, three trimethylated, and one tetramethylated derivative were generated (**Table 1**). In addition, three carboxymethylated compounds were detected. The path that led to these results and the identification of the monomethylated derivatives is described below.

Table 1: LC-DAD-ESI-IMS-qTOF results of the synthesis reaction of methylated cyanidin-3-*O*-glucoside.

Compound Number	$[M]^+$ (*m/z*) Observed	$[M]^+$ (*m/z*) Calculated	Mass Error (mDa)	RT (min)	CCS Value $[M]^+$ (Å²)	Fragment Ions (*m/z*)	Yield (%) [a]	λ_{max} (nm)
reference substances								
cyanidin-3-*O*-glucoside							21 [b]	
	449.1078	449.1078	±0.0	3.42	202.8 ± 0.4	287.0551		518
peonidin-3-*O*-glucoside								
	463.1234	463.1235	−0.1	4.19	207.0 ± 0.3	286.0471, 301.0707		519
methylated compounds								
monomethylated cyanidin-3-*O*-glucoside							8	
1	463.1230	463.1235	−0.5	3.95	207.2 ± 0.2	286.0470, 301.0707		520
2	463.1228	463.1235	−0.7	4.16	208.1 ± 0.3	286.0473, 301.0704		517
3	463.1231	463.1235	−0.4	4.36	207.3 ± 0.3	286.0471, 301.0706		517
dimethylated cyanidin-3-*O*-glucoside							3	
6	477.1386	477.1391	−0.5	4.60	213.5 ± 0.4	300.0624, 315.0861		516
7	477.1397	477.1391	+0.6	4.99	212.9 ± 0.5	315.0862		518
8	477.1390	477.1391	−0.1	5.17	212.3 ± 0.1	300.0638, 315.0858		516
trimethylated cyanidin-3-*O*-glucoside							1	
9	491.1547	491.1548	−0.1	5.38	218.3 ± 0.2	329.1019		520
10	491.1543	491.1548	−0.5	5.97	216.5 ± 0.4	329.1026		518
11	491.1548	491.1548	±0.0	6.08	218.3 ± 0.3	329.1019		517
tetramethylated cyanidin-3-*O*-glucoside							1	
12	505.1686	505.1704	−1.8	6.97	223.3 ± 0.2	343.1201		518
carboxymethylated cyanidin-3-*O*-glucoside							4	
13	507.1132	507.1133	−0.1	4.45	215.7 ± 0.3	287.0544		515
14	507.1134	507.1133	+0.1	4.56	212.2 ± 0.3	287.0549		515
15	507.1134	507.1133	+0.1	5.00	212.8 ± 0.6	287.0551		521

[a] expressed as peonidin-3-*O*-glucoside equivalents [b] remaining after reaction.

2.1 First Observations

Simple hemisyntheses of monomethylated quercetin derivatives with DMS take between 4 and 24 h, depending on the solvents and bases used.[20] The synthesis of monomethylated flavonoids with DMC requires 24 to 36 h according to the literature.[21] In the first experiments with 10 mg of anthocyanin extract, 2 mL DMC, and 0.3 mmol DBU at 90 °C, the compounds that were fully methylated could already be identified as the main products after 22 h. The samples contained methylated cyanidin-3-*O*-glucoside at a mass to charge ratio (*m/z*) 505 and 563 ($[M]^+$). The former is the derivative that is fully methylated at the aglycone, whereas the latter is carboxymethylated in addition to permethylation. No monomethylated derivatives were detectable in the samples examined. For this reason, it was decided to shorten the reaction time to stop the reaction at the point where compounds with a lower degree of methylation, such as monomethylated molecules at *m/z* 463 ($[M]^+$) and dimethylated molecules at *m/z* 477 ($[M]^+$) are formed. The optimal reaction time, temperature, and quantity of the catalyst for the formation of noncarboxymethylated monomethylated cyanidin-3-*O*-glucoside derivatives can be determined with the help of a randomized, central composite design.

2.2 Experimental Design

In the synthesis presented here, it is presumed that the nonnucleophilic base, DBU, deprotonates cyanidin-3-*O*-glucoside. In addition, DBU reacts with DMC via the preferred carboxymethylation mechanism to generate a more activated methylating agent. This reacts in the next step with the deprotonated cyanidin-3-*O*-glucoside to form the desired methylated compound while the catalyst is regenerated. The resulting methoxycarbonate anion decays rapidly to CO_2, and, with a proton, to methanol. It becomes clear that the DBU added is only a catalyst. The mechanism was previously described,[13,19] but still needs to be elucidated in more detail. Since anthocyanins have a pH-dependent structure, this certainly plays a major role in the reaction. It is expected that the flavylium cation form, which is present in acidic environments, is more stable and less reactive than the deprotonated form.

Figure 1 shows the relationship between the yield of synthesized compounds at 90 °C for six hours and the volume of catalyst.

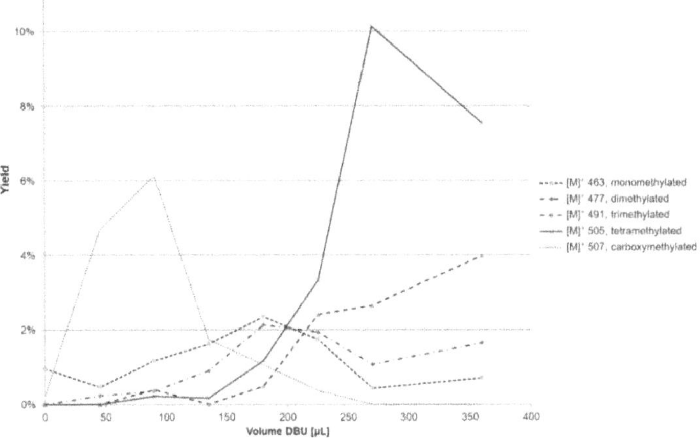

Figure 1: Yield of the synthesis products of cyanidin-3-*O*-glucoside with 2 mL DMC at 90 °C for 6 h as a function of the volume of catalyst DBU.

The tri- and tetramethylated compounds are formed only at higher amounts of DBU, while the amount of mono- and dimethylated derivatives decreases at this point. It should be mentioned that small amounts of peonidin-3-*O*-glucoside were sometimes detected in the blackberry extract used as the starting material. A consideration of the undesired side reaction of carboxymethylation shows that these reactions can be controlled by the concentration of DBU. Without catalyst, the product at *m/z* 507 ($[M]^+$) is not formed, but with 0.6 mmol DBU, its yield is over 6%. This small amount of DBU is immediately consumed as a base and cannot be used to generate the methylation agent. In this way, the anthocyanin reacts directly with DMC, so carboxymethylation is preferred. This consumption of DBU is also described in the literature.[19] With increasing catalyst quantities, a decrease in carboxymethylated cyanidin-3-*O*-glucoside can be observed. With more than 1.2 mmol DBU, the concentration of carboxymethylated cyanidin-3-*O*-glucoside is no longer significant. The result is plausible since DBU reacts as a base and creates the methylating agent from DMC, and thus, only the methyl group is transferred to the cyanidin-3-*O*-glucoside. This pathway also produces CO_2 as the driving force of the reaction. These assumptions could be confirmed in tests with different concentrations of the starting compounds but the same amount of catalyst. Increasing the amount of methylating agent only promotes the side reaction of carboxymethylation. In one of the reactions investigated, a yield greater than 8% was achieved. From the synthesis of *p*-cresol to 4-methyl anisole described in the literature, it can be concluded that a substrate concentration that is too high slows down the reaction kinetics. The reason for this effect is the possibility of the substrate interfering with the formation of the phenolate ion, which is necessary for the transfer of the methyl group, by forming hydrogen bonds between them.[22]

A similar connection is also likely for anthocyanins. With 1 mL of DMC but the same anthocyanin to catalyst ratio, the yield decreases compared to the optimized conditions with 2 mL. Thus, it may be concluded that in case of an excess of methylating agent and a too low concentration of the catalyst, the direct contact of DMC with the anthocyanin leads to carboxymethylation of the sugar residue. The experimental design also confirmed the assumption that carboxymethylation is favored below 90 °C. The experiment resulted in more than one optimal method for maximum yield of monomethylated compounds, while at the same time minimizing the other products. However, since the shortest possible reaction time at high temperatures and high pH values is desirable for anthocyanins, the following conditions were considered optimal: 95 °C for 90 min with 2.95 mmol of DBU. Under these conditions, a total yield of 8% monomethylated cyanidin-3-*O*-glucoside and 4% carboxymethylated cyanidin-3-*O*-glucoside was achieved. After the reaction, approximately 21% of cyanidin-3-*O*-glucoside remained unreacted. Approximately 40% of the compounds present after the reaction could be characterized (**Table 1**). Among the remaining 60%, the degradation products of cyanidin-3-*O*-glucoside, phloroglucine aldehyde, protocatechuic acid, and their methylated derivatives, were detected in the synthesis mixture. Whether these compounds are degradation products of methylated cyanidin-3-*O*-glucoside derivatives, or whether the methylation of these compounds took place after the degradation of cyanidin-3-*O*-glucoside, was not investigated.

2.3 Increase in Solvation of Anthocyanins

During the experiments, it was observed that although the solution took on a red color, some freeze-dried anthocyanin particles were still visible as a suspension in the DMC. This observation is also mentioned in the literature on the methylation of quercetin, where the yield after 72 h is only 30%.[21] To promote the conversion of anthocyanins, in addition to controlling the reaction by adjusting temperature, time, and catalyst concentration, measures to increase the solubility were considered. With this aim, experiments with the application of ultrasound and different solvents were used.

2.3.1 Ultrasound

The overall influence of ultrasound on the synthesis was small. Compared to the synthesis under the optimized conditions, neither an improvement nor a deterioration was observed. It has been reported that the yield of the *O*-alkylation of 5-hydroxy-4-oxo-4H-chromene-2-carboxylic acid could be increased from 28% to 97% by ultrasound.[23] To achieve this, the authors performed 90-min continuous ultrasound treatment with an ultrasonic sonotrode at a frequency of 20 kHz while the system continued to be externally heated. Such an experimental setup could not be realized for the synthesis of methylated cyanidin-3-*O*-glucoside derivatives at the optimal temperature at 95 °C because the sonotrode (Hielscher Ultrasonics GmbH, UP200St) can be used only up to temperatures of approximately 70 °C. Furthermore, ultrasonic treatment during chemical synthesis might result in significant degradation of the products. The mechanisms associated with ultrasound are radical reactions.[24] Those types of reactions, against which anthocyanins may have a stabilizing effect due to their chemical properties, may then lead to countless isomeric polymeric compounds that are difficult to detect.[18]

2.3.2 Solvents

Another approach to obtain more initially dissolved cyanidin-3-*O*-glucoside molecules was the addition of different solvents. The synthesis followed the procedure mentioned under Section 3.2, whereby the sample substance was dissolved in 200 µL solvent such as methanol, DMF, dimethyl sulfoxide (DMSO), water, or pyridine before the addition of DMC and DBU and stirred vigorously. None of the selected solvents were able to increase the yield of monomethylated anthocyanins compared to a reaction without the addition of solvents. In fact, the yield actually decreased. The yield of the higher methylated cyanidin-3-*O*-glucoside derivatives did not exceed 2%. Although the solutions were perceived as more homogeneous before and after the reaction, the yield could not be improved. Although to our knowledge there are no reports of a reaction with the abovementioned solvents except water, where hydrolysis of DMC may occur, such reactions cannot be excluded. It is also possible that a solvent layer forms around the cyanidin-3-*O*-glucoside, and the DMC is unable to react with it. The loss of cyanidin-3-*O*-glucoside increased for all solvents: only between 2 and 4% of the initial amount could be detected without producing more methylated compounds. A plausible explanation for this observation might be an improved heat transfer to the cyanidin-3-*O*-glucoside due to the increased solubility, and thus increased surface and contact possibility. When water is used as the solvent, it is possible that a water molecule reacts with the anthocyanin to form the colorless chromenol, which would no longer be detectable at 520 nm.[8] It should also be taken into account that nucleophilicity is strongly dependent on the solvent. Although DMSO and DMF should not have any influence on the nucleophilicity, as they are not able to form hydrogen bridge bonds, water can have an influence even in small quantities, as it can build hydrate shells around the cyanidin-3-*O*-glucoside hydroxyl groups. The method presented may be further optimized using protective groups, as employed by Bouktaib et al.[5], or the use of phase transfer catalysts, as proposed by Tundo et al.[13].

2.4 Identification

Identification of the synthesized compounds was first carried out with ultra-violet (UV)/Vis detection at 520 nm, since anthocyanins have an absorption maximum in this region, and then by mass spectrometric analysis of the molecular and fragment ions using ultra-high-performance liquid chromatography coupled to electrospray ionization multistage mass spectrometry (UHPLC-ESI-LIT-MS^n). Only those compounds that showed a loss of glucose ($[M - 162]^+$) in the MS^2 spectrum followed by the loss of a methyl group ($[M - 162 - 15]^+$) in the MS^3 spectrum were identified as methylated anthocyanins. To determine the accurate mass and collision cross section (CCS) values, LC-IMS-qTOF-MS was used. **Table 1** gives a summary of the identification characteristics of cyanidin-3-*O*-glucoside and its derivatives at their respective degrees of methylation.

2.4.1 Retention Times (RT)

Regarding the elution order in **Figure 2**, it is evident that methylated cyanidin-3-*O*-glucoside derivatives are, as expected, less polar. The higher the degree of methylation, the later the compounds are eluted. It can also be observed that the retention time of peak **2** is in accordance with that of the reference compound peonidin-3-*O*-glucoside (3'-OH methylated). Another methylated cyanidin-3-*O*-glucoside which may be used as reference is described in the literature as isopeonidin-3-*O*-glucoside (4'-OH methylated). It is formed as a metabolite of human and porcine enzymes.[1,7] Considering only those reports of authors who used acetonitrile (ACN) and water as eluents, the 3'-OH methylated compounds eluted earlier than the 4'-OH. Regarding the elution order of methylated quercetin or catechin derivatives, the literature indicates also that 3'-OH is eluted before 4'-OH, and 5-OH earlier than 7-OH.[25-27] Although the structure is different from the anthocyanins, it may be an orientation in elucidation.

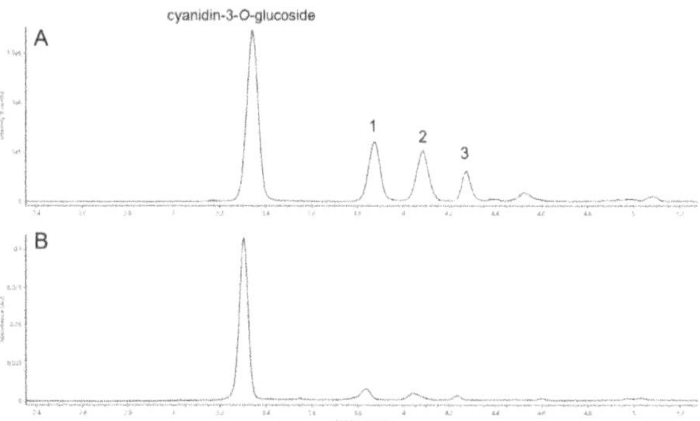

Figure 2: UHPLC-MS chromatograms of the reaction of cyanidin-3-*O*-glucoside with 2 mL DMC and 3.01 mmol DBU. (A), EIC of the *m/z* 449 ($[M]^+$, cyanidin-3-*O*-glucoside) and the *m/z* 463 ($[M]^+$, monomethylated cyanidin-3-*O*-glucoside, peaks 1–3). (B), UHPLC-DAD chromatogram extracted at 520 nm.

2.4.2 UV Spectra

Considering shifts in UV spectra after conjugation can provide hints about the position.[28,29] Unfortunately, the absorption maxima of the methylated derivatives are not significantly lower than that of cyanidin-3-*O*-glucoside. The absorption maxima of the monomethylated compounds and of cyanidin-3-*O*-glucoside are approximately 519 nm. Considering the ratio of $\lambda_{440\ nm}$ to λ_{max}, differences become more noticeable. It can be seen that with 40% peak **2** and **3** have a different ratio than peak **1** with 33%, and the reference substances cyanidin-3-*O*-glucoside, peonidin-3-*O*-glucoside, and 7-*O*-methylcyanidin-3-*O*-galactoside have ratios of approximately 31%. The observations described in the literature, where methylation of cyanidin at the 3'-OH position almost doubles the ratio, could not be confirmed. However, the solvent used was methanol and not ACN.[30] Through the sulfation of cyanidin (unpublished results), the slightest change in this ratio was found at the B ring. If the UV shifts in the two reactions are similar, it can be assumed that peak **1** is derivatized at the B ring and peaks **2** and **3** at the A ring.

2.4.3 Mass Spectrometry

As the MS^3 spectra of the monomethylated derivatives (*m/z* 463; $[M]^+$) represent the fragmentation behavior of the aglycone (*m/z* 301; $[M - 162]^+$), these were used to determine their substitution pattern. The fragment at *m/z* 273 is characteristic of methylated cyanidin and marks the loss of CO of the aglycone ($[M - 162 - 28]^+$). In the same way, the fragment at *m/z* 259 results from the additional loss of the methyl group (*m/z* 287; $[M - 162 - 15 - 28]^+$). All these fragments can also be found in their homolytic cleavage product form.[31] **Figure 3** shows fragments that can be identified as characteristic of the A ring (*m/z* 150, 149, 139) and the B ring (*m/z* 151, 121, 109).[11,32] A simple consideration leads to the conclusion that depending on the substitution, characteristic fragments a mass increment of 14 u are detectable in the spectra of the monomethylated cyanidin-3-*O*-glucoside molecules. In the case of methylation of cyanidin-3-*O*-glucoside, however, this consideration cannot be applied. Since the methyl group is such a small fragment, some of the masses may also belong to other fragments. For example, the fragment at *m/z* 137 is described as a characteristic $^{0,2}B^{+}$-fragment.[32] Nevertheless, this mass may also belong to a methylated fragment of the *m/z* of $[^{1,2}B]^{+}$ (*m/z* 123 + 14) or to the $^{1,3}A^{+}$-fragment.[25] Other examples are the fragment at *m/z* 185 (cyanidin: $[M - H_2O - CO - CO - CO]^{+}$), which may also belong to a methylated fragment of cyanidin ($[M + CH_3 - H_2O - CO - CO - C_2H_2O]^{+}$ (171 + 14)), the fragment at *m/z* 161 (cyanidin: $[M - CO - CO - C_2H_2O - CO]^{+}$ or methylated cyanidin: $[^{1,4}B + CH_3 - H_2O]^{+}$), and the fragment at *m/z* 123 (cyanidin: $[^{1,2}B]^{+}$ or methylated cyanidin: $[^{0,2}B + CH_3 - CO]^{+}$ (109 + 14)).[25,32]

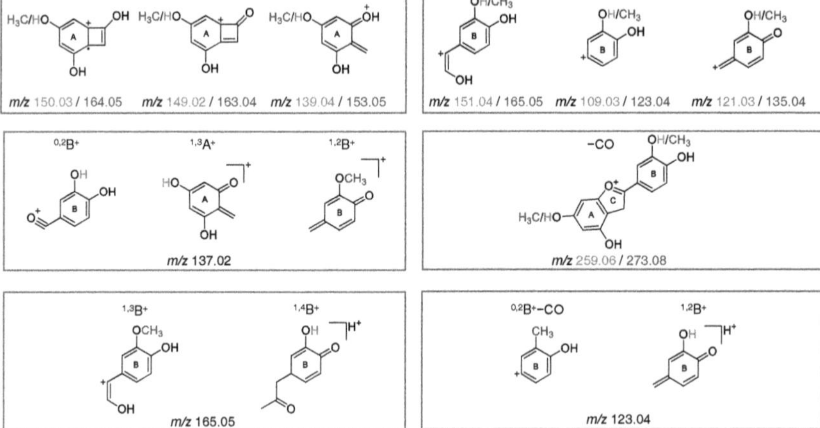

Figure 3: Postulated fragmentation scheme for cyanidin derivatives methylated on the A or B ring, based on the proposals of Barnes and Schug[32] and González-Manzano et al.[25].

2.4.4 Ion Mobility Spectrometry

Over the drift time of the substances, CCS values can be calculated. These depend on the instrument settings, the *m/z*, and the structure and configuration of the molecule. Based on the retention time, it might be concluded that peak **2** is peonidin-3-*O*-glucoside. However, at 208.1 ± 0.3 Å², the CCS value of peak **2** deviates too much from the CCS value of the reference with

207.0 ± 0.3 Å^2. Thus, the possibility that the cyanidin derivative methylated at the 3'-OH position is among the three detected can be cautiously excluded.

2.4.5 Views on Reactivity and Order of Methylation

As the ratios of the three monomethylated peaks to each other are 43:39:18, only 3 methylated derivatives were formed in significant quantities. Thus, the question remains regarding which OH group of cyanidin-3-*O*-glucoside is least reactive. In the case of quercetin, the 5-hydroxyl is by far the least acidic,[25] whereas in cyanidin-3-*O*-glucoside, this position is the one with the highest pk_a after 7-OH.[33] The pk_a values for the OH groups in catechin are almost equal at approximately 9.[25] Therefore, no regioselectivity in the methylation reaction of catechin is expected.[25] However, the lowest amount of catechin monomethylated is observed in position 5, with slightly more monomethylated in 7. Additionally, it is described that catechin is preferentially methylated at 3' and 4'-OH,[27] which may be due to a better accessibility. In the past a methylation order that may be dependent on the acidity of the single OH groups of quercetin was observed, where the most acidic group is methylated first.[34] This might be used to identify the methylation position in anthocyanins, as their OH groups also show quite different acidities. According to this approach, the three most abundant resulting derivatives of cyanidin-3-*O*-glucoside are 7, 5, and 4'-OH. However, it should be noted that the conditions under which the work was carried out were completely different.

2.4.6 Hydrolysis

To obtain more precise information about the aglycones, a sample was hydrolyzed with hydrochloric acid, as the hydrolysis allows the aglycone to be measured as MS^2 rather than as MS^3. The hydrolyzed sample was measured with both LC-ESI-LIT-MS^n and LC-IMS-qTOF-MS and compared with a peonidin standard and a hydrolyzed sample of Brazilian pepper extract. As Brazilian pepper contains 7-*O*-methylcyanidin galactoside,[28] a comparison of the glycosides was not sufficient for identification. The authors succeeded in distinguishing peonidin and 7-*O*-methylcyanidin based on characteristic fragments for the A ring and the B ring. **Figure 4** shows an extracted ion chromatogram (EIC) of the $[M]^+$, *m/z* 301 of a hydrolyzed cyanidin-3-*O*-glucoside sample after methylation compared to a hydrolyzed sample of Brazilian pepper and a peonidin reference.

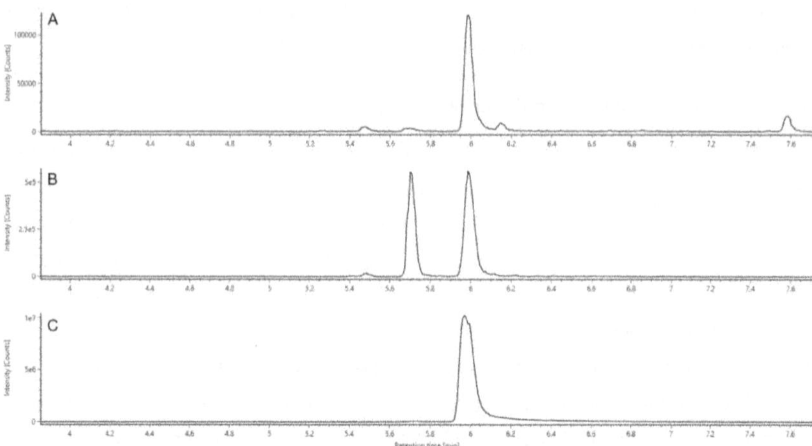

Figure 4: Extracted ion chromatogram of the *m/z* 301 ([M]$^+$, methylated cyanidin) of the methylation of cyanidin-3-*O*-glucoside after hydrolysis (B) compared to a hydrolyzed sample of Brazilian pepper (A) and a peonidin reference (C).

Table 2 provides information about the aglycones found. In comparison with the reference substances, it can be seen from peak **18** that peonidin and 7-*O*-methylcyanidin have the same retention time. The observation that peonidin and 7-*O*-methylcyanidin have similar chromatographic behavior was already described before.[30] The two substances have almost the same Rf value in TLC. Fortunately, IMS is more accurate as an identification technique in this instance. The CCS values allow us to exclude that the detected peak **18** is peonidin because their values differ from each other by 160.3 ± 0.5 Å^2 for peak **18** and 159.1 ± 0.1 Å^2 for peonidin. However, the value of 7-*O*-methylcyanidin at 160.3 ± 0.5 Å^2 is identical to that of peak **18**. Assuming that glucose does not change the elution order, it can be cautiously concluded that peak **3** is 7-*O*-methylcyanidin-3-*O*-glucoside.

Table 2: LC-DAD-ESI-IMS-qTOF results of the synthesis reaction of methylated cyanidin-3-*O*-glucoside after acidic hydrolysis.

Compound Number	[M]$^+$ (*m/z*) Observed	[M]$^+$ (*m/z*) Calculated	Mass Error (mDa)	RT (min)	CCS Value [M]$^+$ (Å^2)	Fragment Ions MS (m/z)	λ_{max} (nm)
reference substances							
Cyanidin							
	287.0548	287.0550	−0.2	4.86	154.7 ± 0.2	213.0545, 231.06493, 241.04919, 269.04454	527
Peonidin							
	301.0711	301.0707	+0.4	5.98	159.1 ± 0.1	201.05474, 229.04958, 258.05195, 286.0471	527
7-*O*-methylcyanidin							
	301.0702	301.0707	−0.5	5.99	160.3 ± 0.5	286.04672, 258.05171	524
methylated compounds							
monomethylated cyanidin							
16	301.0706	301.0707	−0.1	5.48	160.6 ± 0.5	258.05262, 269.04454, 286.04732	n.d.
17	301.0704	301.0707	−0.3	5.71	159.4 ± 0.2	258.05262, 286.04732	528
18	301.0701	301.0707	−0.6	5.99	160.3 ± 0.5	201.05407, 213.05477, 229.04907, 258.05165, 286.04683	524

				dimethylated cyanidin			
19	315.0860	315.0863	−0.3	6.67	164.4 ± 0.3	272.06712, 300.06231	527
20	315.0860	315.0863	−0.3	6.74	164.7 ± 0.6	257.04412, 201.05023, 272.06728, 301.06719,	528
21	315.0858	315.0863	−0.5	6.97	164.9 ± 0.3	257.04467, 271.06016, 300.06297	523
				trimethylated cyanidin [a]			
22	329.1011	329.1020	−0.9	7.78	170.0 ± 0.4	314.07906, 286.08402, 271.06265, 257.08148	526
23	329.1017	329.1020	−0.3	7.94	170.0 ± 0.2	314.07861, 286.08251, 271.06018, 257.08116	524
24	329.1014	329.1020	−0.6	8.09	170.9 ± 0.4	313.07039, 314.0769, 285.07596, 268.07533	524
				tetramethylated cyanidin			
25	343.1177	343.1176	+0.1	8.80	176.8 ± 0.2	327.08619, 312.06264, 299.09129, 285.07529	522

[a] one probably resulted from a carboxymethylated and trimethylated cyanidin-3-*O*-glucoside.

Regarding the absorption maxima at approximately 520 nm, the tendency to undergo a hypsochromic shift as compared to cyanidin is more pronounced for the trimethylated derivatives and the tetramethylated derivative than for mono- and dimethylated cyanidin derivatives. After hydrolysis, three monomethylated compounds can be found. Peak **16** is the smallest, and the ratio of the three is 1:45:54.

An explanation for only two large peaks remaining after hydrolysis might be the coelution of two substances. On the other hand, one of the three compounds may have been decomposed by hydrolysis. This would mean that one of the methyl group positions might have a destabilizing effect on the cyanidin-3-*O*-glucoside molecule. It has been described that methylation in the B ring decreases stability in neutral media.[35] However, opposite observations under acidic conditions and high temperatures, showing a stabilizing effect of the methyl group of peonidin are also described.[36] Concerning the other OH groups, no report could be found.

3 Materials and methods

3.1 Chemicals

LC-MS-grade water, acetonitrile, and formic acid were purchased from ChemSolute (Renningen, Germany). Methanol was obtained from VWR (Darmstadt, Germany). Dimethyl carbonate (DMC), dimethyl sulfoxide (DMSO), and 1,8-diazabicyclo[5.4.0]undec7.ene (DBU) were obtained from Carl Roth (Karlsruhe, Germany). *N,N*-dimethyl formamide (DMF) was purchased from Fisher Scientific (Waltham, Massachusetts). Peonidin and peonidin-3-*O*-glucoside chloride were obtained from Phytoplan (Heidelberg, Germany). An anthocyanin extract containing primarily cyanidin-3-*O*-glucoside (**Scheme 2**) was obtained as described before.[9]

R^1	R^2	R^3	R^4	compound
H	H	H	H	cyanidin-3-*O*-glucoside
CH_3	H	H	H	peonidin-3-*O*-glucoside
H	CH_3	H	H	isopeonidin-3-*O*-glucoside
H	H	CH_3	H	5-*O*-methylcyanidin-3-*O*-glucoside
H	H	H	CH_3	7-*O*-methylcyanidin-3-*O*-glucoside

Scheme 2: Structure of cyanidin-3-*O*-glucoside and its methylated derivatives.

3.2 Hemisynthesis

Ten milligrams of the above-described anthocyanin extract containing approximately 20 µmol cyanidin-3-*O*-glucoside was reacted with 2 mL dimethyl carbonate, which served both as the reactant and the solvent. The reaction was catalyzed by 440 µL (2.95 mmol) DBU at 95 °C for 90 min. It was conducted in a 5 mL reaction vial with cap and septum in a heating unit under constant stirring. This approach was based on a method by Bernini et al.[21] and was optimized by a randomized, central composite design. The investigation was supported by the software Design Expert, which is suitable for response surface methods. A range of 60 °C to 100 °C was considered for the temperature, and a range of 30 to 480 min was used for the reaction time. The volumes of catalyst were chosen between 130 µL and 440 µL. The central point of the experimental design was thus 80 °C, 255 min, and 285 µL (1.91 mmol) DBU. This central point was repeated five times.

Sample Preparation

To the samples obtained through synthesis, 1 mL of methanol was added to remove excessive DMC azeotropes under nitrogen flow. The residue was dissolved in a mixture of acetonitrile, water, and formic acid (18.5/78.5/3; *v/v/v*) and adjusted to a volume of 5 mL. After further dilution (1:2 or 1:10) and microfiltration (regenerated cellulose, 0.2 µm), the samples were analyzed using liquid chromatography coupled with ion trap (LC-LIT-MS^n) and ion mobility quadrupole time of flight (LC-IMS-QTOF-MS) multistage mass spectrometry. For further identification purposes, the glucose moiety of the resulting reaction products was removed through hydrolysis with hydrochloric acid for 90 min at 90 °C. This procedure allowed the comparison with an anthocyanin-rich extract of Brazilian pepper (*Schinus terebinthifolius*), which was treated in the same way.

3.3 LC-MS Analysis

UHPLC analysis of the reaction products was performed on an Acquity UPLC I-Class system (Waters, Milford, MA, USA). The apparatus consisted of a sample manager cooled at 10 °C, a binary pump, a column oven, and a diode array detector scanning from 250 to 650 nm. The column oven temperature was set at 40 °C. An Acquity HSS-T3 RP18 column (150 mm x 2.1 mm; 1.8 µm particle size) was combined with a precolumn (Acquity UPLC HSS T3 VanGuard, 100 Å, 2.1 mm x 5 mm, 1.8 µm), both from Waters (Milford, MA, USA). The separation was performed with water (A) and acetonitrile (B) as eluents, both acidified with 3% (*v* + *v*) formic acid. The flow rate was set at 0.4 mL/min. Analyses of the methylated cyanidin-3-*O*-glucoside derivatives were carried out with linear gradient conditions from 7% B to 25% B for 5 min, then to 50% B for 5 min. For the methylated cyanidin derivatives, the linear gradient ranged from 7% B to 25% B for 5 min, then to 70% B for 9 min. The injection volume was 5 µL. For MS analysis, the UHPLC was coupled with an LTQ-XL ion trap mass spectrometer (Thermo Scientific, Inc., Waltham, MA, USA). This was equipped with an electrospray interface operating in positive ion mode. Ion mass spectra were recorded in the range of *m/z* 245–1000. The source voltage was kept at 4 kV at a current of 100 µA, and the tube lens was adjusted to 55 V. The capillary temperature was set at 325 °C with a spray voltage of 4 V. The sheath, auxiliary, and sweep gas was nitrogen at a flow of 70, 10, and 1 arb, respectively. Three consecutive scans were conducted: a full mass scan, an MS/MS scan of the most abundant ion from the first scan using a normalized collision energy (CE) of 65%, and an MS^3 of the most abundant ion in MS^2 with a CE of 65%.

In addition, multiple reaction monitoring (MRM) measurements were performed on the fragment of the monomethylated aglycone (*m/z* at 301, $[M]^+$) with different collision energies ranging from 10 to 200%. The Xcalibur (2.2SP1.48, Thermo Scientific, Inc., Waltham, MA, USA) was used to evaluate the data. For ion mobility spectrometry measurements, the UHPLC was connected to a Vion IMS QTOF mass spectrometer (Waters, MA, USA). The ionization mode was positive. The capillary voltage was 0.5 kV, the source temperature was 120 °C, the cone voltage was 40 V, the desolvation gas temperature was 550 °C, and the desolvation gas flow was 1200 L/h. Automatic lock correction was conducted every 5 min with leucine-enkephalin as the lock mass at a concentration of 100 pg/µL. The drift gas was nitrogen, and the MS mode was high definition with a low collision energy of 6 eV and a high collision energy ramp of 20–40 eV. Data were acquired and processed using UNIFI v1.9.2.045 (Waters, Milford, MA, USA).

4 Conclusion

With this work, a foundation for the green synthesis of methylated anthocyanin metabolites is provided. It was possible to maximize the synthesis of the desired monomethylated derivatives and minimize undesired side products. Cautious identification of 4'-, 5-, and 7-OH derivatives was possible. Precise identification after isolation would be the next step to use the synthesized components as reference substances. Nuclear magnetic resonance (NMR) spectroscopy is the most promising approach here, as other spectroscopic methods reach their limits. In addition, the methylation of other anthocyanins should be included to investigate whether the aglycone or the sugar moiety have an influence on the synthesis.

Author Contributions: Conceptualization, S.S., M.P. and A.S.; methodology, S.S. and T.B.; validation, S.S.; investigation, S.S. and T.B.; interpretation, S.S., T.B., M.P. and A.S., resources, A.S.; writing—original draft preparation, S.S.; writing—review and editing, T.B., M.P. and A.S.; supervision, M.P. and A.S.; project administration, M.P.; All authors have read and agreed to the published version of the manuscript.

Funding: This research received no external funding.

Data Availability Statement: Data is contained within the article.

Conflicts of Interest: The authors declare no conflict of interest.

Sample Availability: Samples of the compounds are not available from the authors.

5 References

(1) Fernandes, I.; Marques, F.; de Freitas, V.; Mateus, N. Antioxidant and antiproliferative properties of methylated metabolites of anthocyanins. Food Chem. 2013, 141, 2923–2933, doi:10.1016/j.foodchem.2013.05.033.

(2) Mazza, G.J. Anthocyanins and heart health. Ann. Ist. Super. Sanita 2007, 43, 369–374.

(3) Wu, X.; Pittman, H.E.; McKay, S.; Prior, R.L. Aglycones and sugar moieties alter anthocyanin absorption and metabolism after berry consumption in weanling pigs. J. Nutr. 2005, 135, 2417–2424, doi:10.1093/jn/135.10.2417.

(4) Kamiloglu, S.; Capanoglu, E.; Grootaert, C.; van Camp, J. Anthocyanin Absorption and Metabolism by Human Intestinal Caco-2 Cells—A Review. Int. J. Mol. Sci. 2015, 16, 21555–21574, doi:10.3390/ijms160921555.

(5) Bouktaib, M.; Lebrun, S.; Atmani, A.; Rolando, C. Hemisynthesis of all the O-monomethylated analogues of quercetin in-cluding the major metabolites, through selective protection of phenolic functions. Tetrahedron 2002, 58, 10001–10009, doi:10.1016/S0040-4020(02)01306-6.

(6) Blount, J.W.; Redan, B.W.; Ferruzzi, M.G.; Reuhs, B.L.; Cooper, B.R.; Harwood, J.S.; Shulaev, V.; Pasinetti, G.; Dixon, R.A. Synthesis and quantitative analysis of plasma-targeted metabolites of catechin and epicatechin. J. Agric. Food Chem. 2015, 63, 2233–2240, doi:10.1021/jf505922b.

(7) Wu, X.; Pittman, H.E.; Prior, R.L. Pelargonidin is absorbed and metabolized differently than cyanidin after marionberry consumption in pigs. J. Nutr. 2004, 134, 2603–2610, doi:10.1093/jn/134.10.2603.

(8) Cruz, L.; Basílio, N.; Mateus, N.; Pina, F.; de Freitas, V. Characterization of kinetic and thermodynamic parameters of cya-nidin-3-glucoside methyl and glucuronyl metabolite conjugates. J. Phys. Chem. B 2015, 119, 2010–2018, doi:10.1021/jp511537e.

(9) Schmitt, S.; Tratzka, S.; Schieber, A.; Passon, M. Hemisynthesis of Anthocyanin Phase II Metabolites by Porcine Liver Enzymes. J. Agric. Food Chem. 2019, 67, 6177–6189, doi:10.1021/acs.jafc.9b01315.

(10) Cruz, L.; Mateus, N.; de Freitas, V. First chemical synthesis report of an anthocyanin metabolite with in vivo occurrence: Cyanidin-4′-O-methyl-3-glucoside. Tetrahedron Lett. 2013, 54, 2865–2869, doi:10.1016/j.tetlet.2013.03.100.

(11) Cren-Olivé, C.; Lebrun, S.; Rolando, C. An efficient synthesis of the four mono methylated isomers of (+)-catechin including the major metabolites and of some 97emethylated and trimethylated analogues through selective protection of the catechol ring. J. Chem. Soc. Perkin Trans. 1 2002, 6, 821–830, doi:10.1039/b107340k.

(12) Kozłowska, J.; Potaniec, B.; Żarowska, B.; Anioł, M. Synthesis and Biological Activity of Novel O-Alkyl Derivatives of Naringenin and Their Oximes. Molecules 2017, 22, 1485, doi:10.3390/molecules22091485.

(13) Tundo, P.; Musolino, M.; Aricò, F. The reactions of dimethyl carbonate and its derivatives. Green Chem. 2018, 20, 28–85, doi:10.1039/C7GC01764B.

(14) Tundo, P.; Memoli, S.; Hérault, D.; Hill, K. Synthesis of methylethers by reaction of alcohols with dimethylcarbonate. Green Chem. 2004, 6, 609–612, doi:10.1039/B412722F.

(15) Lee, Y.; Shimizu, I. Convenient O-Methylation of Phenols with Dimethyl Carbonate. Synlett 1998, 1998, 1063–1064, doi:10.1055/s-1998-1893.

(16) Ouk, S.; Thiebaud, S.; Borredon, E.; Legars, P.; Lecomte, L. O-Methylation of phenolic compounds with dimethyl carbonate under solid/liquid phase transfer system. Tetrahedron Lett. 2002, 43, 2661–2663, doi:10.1016/S0040-4039(02)00201-0.

(17) Ouk, S.; Thiébaud, S.; Borredon, E.; Le Gars, P. High performance method for O-methylation of phenol with dimethyl car-bonate. Appl. Catal. A 2003, 241, 227–233, doi:10.1016/S0926-860X(02)00467-2.

(18) Rein, M. Copigmentation Reactions and Color Stability of Berry Anthocyanins; University of Helsinki: Helsinki, Finland, 2005; ISBN 952-10-2293-0.

(19) Shieh, W.C.; Dell, S.; Repic, O. 1,8-Diazabicyclo5.4.0undec-7-ene (DBU) and microwave-accelerated green chemistry in methylation of phenols, indoles, and benzimidazoles with dimethyl carbonate. Org. Lett. 2001, 3, 4279–4281, doi:10.1021/ol016949n.

(20) Tatsuzaki, J.; Ohwada, T.; Otani, Y.; Inagi, R.; Ishikawa, T. A simple and effective preparation of quercetin pentamethyl ether from quercetin. Beilstein J. Org. Chem. 2018, 14, 3112–3121, doi:10.3762/bjoc.14.291.

(21) Bernini, R.; Crisante, F.; Ginnasi, M.C. A convenient and safe O-methylation of flavonoids with dimethyl carbonate (DMC). Molecules 2011, 16, 1418–1425, doi:10.3390/molecules16021418.

(22) Ouk, S.; Thiébaud, S.; Borredon, E.; Le Gars, P. Dimethyl carbonate and phenols to alkyl aryl ethers via clean synthesis. Green Chem. 2002, 4, 431–435, doi:10.1039/B203353B.

(23) Mason, T.J. Ultrasound in synthetic organic chemistry. Chem. Soc. Rev. 1997, 26, 443, doi:10.1039/CS9972600443.

(24) Riesz, P.; Berdahl, D.; Christman, C.L. Free radical generation by ultrasound in aqueous and nonaqueous solutions. Environ. Health Perspect. 1985, 64, 233–252, doi:10.1289/ehp.8564233.

(25) González-Manzano, S.; González-Paramás, A.; Santos-Buelga, C.; Dueñas, M. Preparation and characterization of catechin sulfates, glucuronides, and methylethers with metabolic interest. J. Agric. Food Chem. 2009, 57, 1231–1238, doi:10.1021/jf803140h.

(26) Cren-Olivé, C.; Déprez, S.; Lebrun, S.; Coddeville, B.; Rolando, C. Characterization of methylation site of monomethylfla-van-3-ols by liquid chromatography/electrospray ionization tandem mass spectrometry. Rapid Commun. Mass Spectrom. 2000, 14, 2312–2319, doi:10.1002/1097-0231(20001215)14:23<2312:AID-RCM160>3.0.CO;2-A.

(27) Donovan, J.L.; Luthria, D.L.; Stremple, P.; Waterhouse, A.L. Analysis of (+)-catechin, (−)-epicatechin and their 3′- and 4′-O-methylated analogs. J. Chromatogr. B Biomed. Sci. Appl. 1999, 726, 277–283, doi:10.1016/S0378-4347(99)00019-5.

(28) Feuereisen, M.M.; Hoppe, J.; Zimmermann, B.F.; Weber, F.; Schulze-Kaysers, N.; Schieber, A. Characterization of phenolic compounds in Brazilian pepper (Schinus terebinthifolius Raddi) exocarp. J. Agric. Food Chem. 2014, 62, 6219–6226, doi:10.1021/jf500977d.

(29) Day, A.J.; Bao, Y.; Morgan, M.R.; Williamson, G. Conjugation position of quercetin glucuronides and effect on biological activity. Free Radic. Biol. Med. 2000, 29, 1234–1243, doi:10.1016/S0891-5849(00)00416-0.

(30) Toki, K.; Saito, N.; Irie, Y.; Tatsuzawa, F.; Shigihara, A.; Honda, T. 7-O-Methylated anthocyanidin glycosides from Catharanthus roseus. Phytochemistry 2008, 69, 1215–1219, doi:10.1016/j.phytochem.2007.11.005.

(31) Hvattum, E.; Ekeberg, D. Study of the collision-induced radical cleavage of flavonoid glycosides using negative electrospray ionization tandem quadrupole mass spectrometry. J. Mass Spectrom. 2003, 38, 43–49, doi:10.1002/jms.398.

(32) Barnes, J.S.; Schug, K.A. Structural characterization of cyanidin-3,5-diglucoside and pelargonidin-3,5-diglucoside anthocya-nins: Multi-dimensional fragmentation pathways using high performance liquid chromatography-electrospray ionization-ion trap-time of flight mass spectrometry. Int. J. Mass Spectrom. 2011, 308, 71–80, doi:10.1016/j.ijms.2011.07.026.

(33) Dangles, O.; Fenger, J.A. The Chemical Reactivity of Anthocyanins and Its Consequences in Food Science and Nutrition. Molecules 2018, 23, 1970, doi:10.3390/molecules23081970.

(34) Rao, K.V.; Owoyale, J.A. Partial methylation of quercetin: Direct synthesis of tamarixetin, ombuin and ayanin. J. Heterocycl. Chem. 1976, 13, 1293–1295, doi:10.1002/jhet.5570130629.

(35) Fleschhut, J.; Kratzer, F.; Rechkemmer, G.; Kulling, S.E. Stability and biotransformation of various dietary anthocyanins in vitro. Eur. J. Nutr. 2006, 45, 7–18, doi:10.1007/s00394-005-0557-8.

(36) Tomaz, I.; Šikuten, I.; Preiner, D.; Andabaka, Ž.; Huzanić, N.; Lesković, M.; Karoglan Kontić, J.; Ašperger, D. Stability of polyphenolic extracts from red grape skins after thermal treatments. Chem. Pap. 2019, 73, 195–203, doi:10.1007/s11696-018-0573-9.

Chapter 5

Concluding remarks

1 Impact of contributions

The sulfation, methylation, glucuronidation of phenolic compounds has been performed for years due to its importance and has been described in detail.[1,2,3,4,5] Each of these publications contributed to the synthesis of further compounds. The synthesis of sulfated, methylated and glucuronidated anthocyanins, however, is rarely reported. This may be due to the fact that quercetin and catechin, for example, are the most described compounds and also the most abundant in nature, but also because anthocyanins are more difficult to handle. With their positive charge they have a special position among the flavonoids. Anthocyanins react differently to the environment such as solvents, temperature, light, and pH value than the other flavonoids and are usually more sensitive. Therefore, only recommendations on how to handle anthocyanins could be derived from the experiments already described. The present work provides more concrete suggestions on how to chemically produce methylated and sulfated anthocyanins. Moreover, first steps for a chemical glucuronidation are introduced.

The method to isolate cyanidin-3-*O*-glucoside is a cost-effective way to get the starting materials that are most needed. In this manner, hemisynthesis is the easier way to produce metabolites than full synthesis, as this comprises many steps, which can be saved in this way.

The experiments with pig liver enzymes not only provide valuable information about the metabolites produced. A further contribution could also be made to elucidate the metabolism in pigs. Since this metabolism is quite similar to the human metabolism,[6] further information can be derived from it, which can help in the research of the complete human metabolism of polyphenols. This will bring us a small step closer to understanding the pathways of the individual metabolites in the body and the reasons why which metabolite is formed where and when.

The metabolites produced via the syntheses described and the data obtained can be used in the future: First of all, the spectrometric, photometric and chromatographic data can be used to build up a comparative database. Thus, physiological samples can be analyzed in the future and with the help of these data, possibly present metabolites can be identified much easier. This database can then be used by a wide variety of research institutions.

Another benefit of the work is the use of the metabolites as reference substances. If actually isolated, pure metabolites can be obtained and stored in a stable manner. They can then be used as references in LC-MS experiments and in cell experiments for further research on bioavailability, absorption, distribution, and overall health effects. Thus, the answer to the question whether the polyphenolic compounds or their metabolites actually have the positive effect in the so-called super foods and fruits is a step closer.

Since ion mobility spectrometry is a relatively new application here, the data obtained can also contribute to the development of a worldwide CCS value database. Since not many values exist yet, the data of this work can contribute to serve as a comparison for other researchers. In addition, further information can be derived from this data, which can be used, for example, in the development of computer programs to calculate CCS values.

2 Remaining challenges and future directions

The metabolites synthesized could partially be identified by measuring the UV/Vis and MS spectra and CCS values. However, these techniques have their limit, and a certainty of 100% and complete identification of all metabolites produced is still pending. This should be achieved by nuclear magnetic resonance spectroscopy. The application of different techniques is possible. Coupling with liquid chromatography has the advantage that prior isolation is not necessary. However, the measuring times are quite short, which leads to a low sensitivity. For an exact measurement, therefore, enough purified material must be available.

Finding an ideal method to isolate the anthocyanin metabolites took up a large part of this work. As a first step to isolate the sulfated metabolites, anion exchange SPE was successfully applied. As described in **chapter 3**, other researchers have already failed in further purification by preparative HPLC. First steps to find a suitable method using preparative HPLC were made in the framework of this work, but still need to be completed in the future. A technique frequently described in synthetic chemistry for product purification is flash chromatography. This technique could initially be used to provide the starting materials much more efficiently. However, the resulting products could also be separated from the remaining starting materials and other reaction products with the right method and on a large scale. In this way, a relatively pure product could be obtained quickly.

Chapter 1 section 3.2.4 describes the hemisynthesis of cyanidin-3-*O*-glucoside glucuronides. The Königs-Knorr synthesis initially delivered promising results but failed to be reproducible. The path chosen then of the imidate method could achieve first positive results. This path should be followed in the future to obtain glucuronides from cyanidin-3-*O*-glucoside and cyanidin.

As the enzymes in the living organism form not only "pure" derivatives of one chemical species, a combination of different reactions is particularly interesting. As can be seen from the experiments with porcine liver and the analysis of urine after blackberry juice consumption in **chapters 2** and **3**, mixed metabolites are most abundant. The most frequently detected metabolites were combinations of methylated and glucuronidated cyanidin-3-*O*-glucoside. Thus, it would be interesting to perform one reaction first and then the other reaction with the isolated product. It would also be conceivable to develop a synthesis in which both reactions take place simultaneously or shortly after each other. This would avoid the time-consuming step of isolation, which is harmful to the sensitive anthocyanin metabolites.

An interesting technique to produce specific derivatives is the use of protective groups. With this, certain hydroxyl groups on the molecule can first be protected by a reaction, so that only a specific group is available for the following derivatization reaction. The challenge here again is the sensitivity of anthocyanins to high pH values and temperatures, since the introduction and removal of protective groups often require such conditions.

In nature there are many more phenolic compounds. Not much is known about their metabolism. Only similar compound classes can be deduced from their ADME. In order to investigate them further, reference substances must be synthesized.

The foundation for a database of phenolic metabolites was laid with this work. Various metabolites of different phenolic compound classes were synthesized and measured. A more extensive synthesis including combined metabolites and further compound classes with a subsequent isolation and stable storage should be the goal for further work.

3 References

(1) Al-Maharik, N.; Botting, N. P. A facile synthesis of isoflavone 7-*O*-glucuronides. *Tetrahedron Lett.* **2006**, *47,* 8703–8706.

(2) Araújo, K. C. F.; M B Costa, E. M. de; Pazini, F.; Valadares, M. C.; Oliveira, V. de Bioconversion of quercetin and rutin and the cytotoxicity activities of the transformed products. *Food Chem. Toxicol.* **2013**, *51,* 93–96.

(3) Barron, D. Recent advances in the chemical synthesis and biological activity of phenolic metabolites. In *Recent advances in polyphenol research;* Daayf, F.; Lattanzio, V., Eds.; Wiley-Blackwell: Oxford, UK, **2008**, 317–358.

(4) Barron, D.; Ibrahim, R. K. Synthesis of flavonoid sulfates: 1. stepwise sulfation of positions 3, 7, and 4 using *N,N'*-dicyclohexylcarbodiimide and tetrabutylammonium hydrogen sulfate. *Tetrahedron* **1987**, *43,* 5197–5202.

(5) Cren-Olivé, C.; Lebrun, S.; Rolando, C. An efficient synthesis of the four mono methylated isomers of (+)-catechin including the major metabolites and of some dimethylated and trimethylated analogues through selective protection of the catechol ring. *J. Chem. Soc., Perkin Trans. 1* **2002,** 821–830.

(6) Miller, E. R.; Ullrey, D. E. The pig as a model for human nutrition. *Annu. Rev. Nutr.* **1987**, *7,* 361–382.

Summary

Polyphenols are considered healthy because they are supposed to protect people from civilization diseases such as cardiovascular diseases, cancer, or diabetes. This may be due to the inhibition of inflammation or their antioxidant effects. However, there is still a lot of research to be done in this direction because their absorption, distribution, metabolism, and excretion depend on many factors such as age, sex, diet, or microbiome. Moreover, it is possible that the metabolites have a different bioactivity than the original substances. Therefore, it is necessary to accurately identify and quantify the anthocyanin metabolites, and for that reference substances are urgently needed. Because they are not commercially available and cannot be isolated from plants or physiological samples their synthesis is indispensable. After an anthocyanin is absorbed, the sugar bond may be cleaved. Then either the glucoside or the aglycone can be methylated by catechol-*O*-methyltransferase, with S-adenosylmethionine serving as the cofactor. Another possibility of metabolism is sulfation, where a sulfonate group is transferred to the molecule with the aid of a sulfotransferase and phosphoadenosine phosphosulfate as cofactor. A third possibility of derivatization is glucuronidation, where a glucuronic acid is transferred by the glucuronosyltransferase of UDP-glucuronic acid. Combinations of these pathways and diverse positions are also likely. In addition, the catabolism by the microbiota should not be ignored. As a result, reference substances are needed to understand the metabolism of anthocyanins. Their synthesis can be achieved in different ways: the most obvious realized in this work are enzymatic or chemical approaches. For the enzymatic way, the enzyme source was porcine liver, which was separated into various fractions that contain different enzymes. Pork liver was chosen because pig metabolism is similar to humans. As a starting material, a blackberry extract was used from which the anthocyanins were extracted. Blackberries contain almost exclusively cyanidin-3-*O*-glucoside. This compound and the cyanidin aglycone were incubated with the porcine liver fractions and corresponding cofactors to obtain the desired metabolites. For methylation, the S9 fraction of liver was chosen. Through the mass spectra and retention time comparison with a reference, the resulting compound could be identified as peonidin-3-*O*-glucoside. Sulfation of the above-mentioned components could also be achieved in this way. Three different sulfates of cyanidin and its 3-*O*-glucoside were detected. The third reaction is glucuronidation, which was conducted with the microsomale fraction of the liver. Three glucuronidated derivatives of cyanidin-3-*O*-glucoside were detected. Although it is not yet possible to determine the exact position using mass spectrometry, the mass spectra of the three compounds differ in the intensity of their fragment ions. If an unambiguous identification has then taken place through e.g., chemical synthesis, an identification can be possible through mass spectrometry. In this way, the cyanidin aglycone could also be glucuronidated. However, as the unambiguous identification via NMR was not yet possible due to the limited amounts obtained, chemical synthesis is urgently necessary. The first reaction implemented is the sulfation of cyanidin or cyanidin glucoside. These compounds react in a one-pot synthesis with a sulfur trioxide-N-triethylamine complex in a solvent for 60 to 90 minutes at 40 to 50 °C. Result is a mixture of different mono- and disulfates. These can then be separated using preparative HPLC and analyzed via NMR. A comparison with real physiological samples containing sulfated metabolites showed the suitability of the synthesized sulfates as reference substances. The methylation of cyanidin glucoside is also possible by chemical means. It was possible to replace the highly toxic standard methylation reagents methyl iodide and dimethyl sulfate with the "green" chemical dimethyl carbonate. By using DBU a catalyst, the temperature was reduced to 90 °C and the reaction time to 1.5 hours. To analyze the results in an advanced way, ion mobility mass spectrometry coupled to a qToF was used. In addition to the different fragment intensities in the mass spectra, IMS provides a further distinguishing feature for isobaric metabolites. From the obtained data it is possible to establish a database, which may be used to identify metabolites in real biological samples. This is a benefit for the untargeted metabolomics and allows further elucidation of the metabolism of anthocyanins and thus the health-promoting effect of anthocyanins.

Zusammenfassung

Polyphenole gelten als gesund, weil sie die Menschen vor Zivilisationskrankheiten wie Herz-Kreislauf-Erkrankungen, Krebs oder Diabetes schützen sollen. Dies kann auf die Hemmung von Entzündungen oder ihre antioxidative Wirkung zurückzuführen sein. Es gibt jedoch noch viel Forschungsbedarf in dieser Richtung, da die Absorption, die Verteilung, der Metabolismus und die Ausscheidung der Polyphenole von vielen Faktoren wie Alter, Geschlecht, Ernährung oder Mikrobiom abhängen. Darüber hinaus ist es möglich, dass die Metabolite der Polyphenole eine andere Bioaktivität haben als die ursprünglichen Substanzen. Daher ist es notwendig, die Anthocyan-Metabolite genau zu identifizieren und zu quantifizieren, wofür Referenzsubstanzen wichtig sind. Sie sind jedoch nicht im Handel erhältlich und können nicht aus Pflanzen oder physiologischen Proben isoliert werden. Daher ist ihre Synthese unerlässlich. Im menschlichen Organismus ist es bei der Aufnahme eines Anthocyans zunächst einmal möglich, dass die Zuckerbindung gespalten wird. Dann kann entweder das Glucosid oder das Aglykon durch die Catechol-*O*-Methyltransferase methyliert werden. Cofaktor in dieser Reaktion ist S-Adenosylmethionin. Eine weitere Möglichkeit des Metabolismus ist die Sulfatierung. Eine Sulfonatgruppe wird mit Hilfe einer Sulfotransferase und Phosphoadenosinphosphosulfat als Cofaktor auf das Molekül übertragen. Eine dritte Möglichkeit der Derivatisierung ist die Glucuronidierung. Hierbei wird eine Glucuronsäure durch die Glucuronosyltransferase der UDP-Glucuronsäure übertragen. Natürlich sind auch Kombinationen dieser unterschiedlichen Reaktionen und unterschiedlichste Positionen möglich. Darüber hinaus darf der Katabolismus durch die Mikrobiota nicht vernachlässigt werden. Es werden also Referenzsubstanzen benötigt, um den Metabolismus der Anthocyane zu verstehen. Ihre Synthese kann auf verschiedene Weise erreicht werden: die offensichtlichsten und in dieser Arbeit realisierten sind enzymatischer oder chemischer Art. Für die enzymatische war die verwendete Enzymquelle Schweineleber, die in verschiedene Fraktionen aufgeteilt wurde, die unterschiedliche Enzyme enthalten. Es wurde Schweineleber gewählt, da der Stoffwechsel von Schweinen dem des Menschen ähnlich ist. Als Ausgangsmaterial wurde ein Brombeerextrakt verwendet, aus dem die Anthocyane extrahiert wurden. Diese enthalten fast ausschließlich Cyanidin-3-*O*-glucosid. Dieses Cyanidin-3-*O*-glucosid und das Cyanidin-Aglykon wurden mit den Schweineleberfraktionen und entsprechenden Cofaktoren inkubiert, um die gewünschten Metaboliten zu erhalten. Für die Methylierung wurde die S9-Fraktion der Leber gewählt. Durch die Massenspektren und den Retentionszeitvergleich mit einer Referenz konnte die resultierende Verbindung als Peonidin-3-*O*-glucosid identifiziert werden. Auch die Sulfatierung der oben genannten Komponenten konnte auf diese Weise erreicht werden. Es wurden drei verschiedene Cyanidinsulfate und Cyanidin-3-*O*-glucosidsulfate nachgewiesen. Die dritte Reaktion ist die Glucuronidierung, die mit der mikrosomalen Fraktion der Leber durchgeführt wurde. Es konnten drei glucuronidierte Derivate von Cyanidin-3-*O*-glucosid nachgewiesen werden. Obwohl es noch nicht möglich ist, die genaue Position der Konjugation mittels Massenspektrometrie zu bestimmen, unterscheiden sich die Massenspektren der drei ersteren Verbindungen in der Intensität ihrer Fragmentionen. Wenn dann eine eindeutige Identifizierung z.B. durch chemische Synthese stattgefunden hat, kann eine Identifizierung durch Massenspektrometrie möglich sein. Da die eindeutige Identifizierung mittels NMR aufgrund der sehr geringen Mengen jedoch noch nicht möglich war, ist die chemische Synthese dringend notwendig. Die erste Reaktion ist die Sulfatierung von Cyanidin oder Cyanidinglucosid. Diese reagieren in einer Eintopf-Synthese mit einem Schwefeltrioxid-N-Triethylamin-Komplex in einem Lösungsmittel 60 bis 90 Minuten lang bei 40 bis 50 °C. Dabei entsteht ein Gemisch aus verschiedenen Mono- und Disulfaten. Diese können dann mittels präparativer HPLC getrennt und mittels NMR analysiert werden. Ein Vergleich mit realen physiologischen Proben, die sulfatierte Metabolite enthalten, zeigte die Eignung der synthetisierten Sulfate als Referenzsubstanzen. Die Methylierung von Cyanidinglucosid ist auch auf chemischem Wege möglich. Es gelang, die hochtoxischen Standard-Methylierungsreagenzien Methyljodid und Dimethylsulfat durch die "grüne" Chemikalie Dimethylcarbonat zu ersetzen. Durch die Verwendung von DBU als Katalysator konnte die Temperatur auf 90 °C und die Reaktionszeit auf 1,5 Stunden reduziert werden. Um die Ergebnisse zu analysieren, wurde die Ionenmobilitäts-Massenspektrometrie gekoppelt an ein qTof verwendet. Zusätzlich zu den unterschiedlichen Fragment-Intensitäten in den Massenspektren liefert das IMS ein weiteres Unterscheidungsmerkmal für isobare Metabolite. Aus diesen Daten ist es möglich, eine Datenbank zu erstellen die zur Identifizierung von Metaboliten in realen biologischen Proben verwendet werden kann. Dies ist ein Vorteil für die nicht-zielgerichtete (untargeted) Metabolomik und ermöglicht eine weitere Aufklärung des Metabolismus von Anthocyanen und damit der gesundheitsfördernden Wirkung von Anthocyanen.

Acknowledgement

This work would not have been possible without the help and support of numerous people. I would like to take this opportunity to thank them.

First, I would like to thank Prof. Andreas Schieber for giving me the opportunity to work on this thesis in his group and for the freedom I had in researching and working on this exciting topic.

I would also like to thank Prof. Wüst for being my second supervisor as well as Prof. Wagner and PD. van Echten-Deckert for their participation in the examination committee.

A huge thank you goes to Dr. Maike Passon for her supervision during the time of the doctoral thesis. I am very grateful for the encouragement, exciting discussions, help, advice, support, delicious cooking evenings and the fact that she was always there for me in the last five years.

I am immensely grateful to have spent my doctoral studies with such a wonderful team. All my colleagues were always helpful and I could count on each and every one of them when it came to technical discussions and problems in the lab, but also when cooking, eating, drinking wine or doing other activities together. I hope these friendships will continue for a long time. I would like to say a special thankyou to some of them: Rita for her open ear, for the great conversations in the morning at breakfast and for her open, positive, helpful and just great manner; Timo, Hannes and Sandra for the time we spent together in “Büro 314”. It was always nice to laugh, snack and chat with you. Thank you for taking me in! The team of “MS-Leidensgenossen”, without you the time at the institute would only have been half as educational and fun. I would like to thank the AG PKB with Maike, Micha, Lisa and the adopted Maria for the many hours together in the MS room, the chocolate puddings afterwards and the great cooking evenings! Lena, for making this happen together. I'm glad we got to know each other and were able to spend this time together. Ingrid for the friendship that has developed and for the great time we had in Malta. Mia, who joined us at just the right time and who brought me the hope I needed.

A special thank goes to my students Sebastian, Christine, Annika, Hendrik, Mona and Till, who have contributed to this work with their diligence, their discussions and their ideas in a valuable and great way.

A very special and big thank also goes to my family, friends and my love Martin for all the support and love and that they encouraged me to never give up.

www.ingramcontent.com/pod-product-compliance
Ingram Content Group UK Ltd.
Pitfield, Milton Keynes, MK11 3LW, UK
UKHW021959190726
13853UKWH00004B/1630

9 783736 975255